Theology, Ethics, and Technology in the Work of Jacques Ellul and Paul Virilio

Theology, Ethics, and Technology in the Work of Jacques Ellul and Paul Virilio

A Nascent Theological Tradition

Michael Morelli

LEXINGTON BOOKS
Lanham • Boulder • New York • London

Published by Lexington Books
An imprint of The Rowman & Littlefield Publishing Group, Inc.
4501 Forbes Boulevard, Suite 200, Lanham, Maryland 20706
www.rowman.com

6 Tinworth Street, London SE11 5AL, United Kingdom

British Library Cataloguing in Publication Information Available

Library of Congress Cataloging-in-Publication Data

Names: Morelli, Michael, 1986–author.
Title: Theology, ethics, and technology in the work of Jacques Ellul and Paul Virilio: a nascent theological tradition / Michael Morelli.
Description: Lanham : Lexington Books, [2021] | Includes bibliographical references and index.
Identifiers: LCCN 2021018622 (print) | LCCN 2021018623 (ebook) | ISBN 9781793625434 (cloth) | ISBN 9781793625441 (epub) Subjects: LCSH: Technology—Philosophy. | Technology—Social aspects. | Ellul, Jacques, 1912-1994. | Virilio, Paul.
Classification: LCC T14 .M58 2021 (print) | LCC T14 (ebook) | DDC 601—dc23
LC record available at https://lccn.loc.gov/2021018622
LC ebook record available at https://lccn.loc.gov/2021018623

For Jayne, Hunter, and Findlay

Contents

Acknowledgments

To provide a full acknowledgment of every person who has contributed to the completion of this book would require a book length document. Whether it is family and friends from our Canadian home, or friends we consider family from our home away from home in Scotland, my wife, two sons, and I have been supported by many people. For the names included here, my appreciation for you exceeds the little I am able to communicate in ink. For the names that do not appear here, I am equally grateful. In both cases, I will make up for the lack of thankful ink with meals and words of gratitude.

I would like to begin by thanking my ecclesial communities in Canada and Scotland. I would not have been inspired to start the research presented in this book if it were not for my time spent living, worshipping, and serving in the churches and communities I have. I am likewise grateful for the academic communities I have studied and worked with in Canada and the United Kingdom. The more I study theology with other people, the more I am convinced that the best theology appears when it is read, discussed, and written in community. A special thank you to my colleagues at Northwest College and Seminary, ACTS Seminaries, and Trinity Western University. It is a privilege to work alongside all of you.

I would also like to thank Lexington Books for their efforts to make this book a reality. Jacob Van Vleet and Michael Gibson, thank you for conspiring to see this book published with Lexington. Trevor Crowell, thank you for Solomonic wisdom and seeing this project through to the end with me, especially as deadlines were graciously adjusted in light of the health challenges I experienced during manuscript revisions.

Jacob Marques-Rollison, thank you for your lucid, precise, and generously given thoughts. Dan Patterson, thank you for your illuminating and clever insights. Kevin Hargaden, thank for your erudite, Vonnegut-esque, wit.

David Lilley, Allen Calhoun, and Kevin O'Farrell, thank you for being great officemates, brilliant colleagues, and close friends. Mike Mawson, thank you for saying in that meeting, "It'll be good to see you here in the fall." Brian Brock, thank for your wisdom and care. You were integral to the completion of this book, and on top of that, my family's time in Aberdeen was richer for having spent time with the Brocks.

Norm and Kathy, thank you for your support and for the "office space" free of charge when all of us were in lockdown. Dad and Mom, thank you for surrounding me with stories and always encouraging me, in word and deed, to live in the world of learning. David and Peter, thank you for our many conversations and friendship.

Hunter and Findlay, you never seem to worry about how my writing is going. You are more concerned about playing with no agenda and one of you once told me my writing is boring. All of this is a gift to me. I hope that one day, when both of you are older, neither of you will find my writing boring. Jayne, this book is as much the product of your effort as it is my own. Thank you for your love, acuity, strength, and humor, all of which never cease to fill me with love and joy.

Introduction

A Nascent Theological Tradition

It is explicable though lamentable that the connections between Jacques Ellul and Paul Virilio do not attract attention from scholars in and beyond the disciplines of theology and ethics. Scholars who are interested in Ellul tend to be uninterested in Virilio. Scholars who are interested in Virilio tend to be uninterested in Ellul. Theologians and theological ethicists tend to have little or no interest in either Ellul or Virilio.[1] Explanations and lamentations notwithstanding, the relationship between Ellul and Virilio reveals a nascent theological tradition. Such is the focus of this book. It locates the seed of a tradition buried in the soil of postwar France and waters that seed until it flowers into something fruitful for scholars, especially theological ethicists.

The tradition I am referring to is one in which the word *technology* is functionally synonymous with the word *modernity*, because at least since the twentieth century, modernity cannot seem to be what it is without technology. It is also a tradition in which the relationship between technology and modernity causes problems that ought to be taken seriously enough to be examined from multiple perspectives, including sociology, philosophy, theology, and ethics. Finally, it is a theological tradition, because the thinkers who comprise it are Christians who blend theology with sociology, philosophy, and ethics to expose and resist the problems produced by often brutal pairings of technology and modernity.

The first chapter situates Ellul and Virilio within the socio-historical context of postwar France. It shows why their lives and work assumed similar trajectories, and eventually, created the theological description I described above. It begins with a survey of the brilliant and brutal scene of life in postwar France to answer the questions: Why did Ellul and Virilio focus on technology in their writing, why did they convert to Christianity, and how did their faith shape the ways they thought and wrote about technology? Once both thinkers are situated in their socio-historical context and these questions are answered, I provide documentary evidence, which shows that Ellul and Virilio were engaged in a scholarly dialogue near the end of Ellul's life. The dialogue was interrupted by Ellul's death in 1994, but only partially so. The chapters that follow the first show how Virilio, the younger scholar by twenty years, continued the dialogue in ways that transformed it into a nascent theological tradition.

The second chapter compares Ellul's sociological-historical method with Virilio's philosophical-phenomenological method to establish how they align and diverge in their work on technology. As I define key methods, terms, and concepts in their work, I show Ellul's primary concern with modern technology is its potential to manipulate language, whereas Virilio's primary concern with modern technology is its potential to manipulate the senses. This will accomplish two things. First, it will capture the prominent and valuable features of Ellul and Virilio's sociological and philosophical work on modern technology. Second, it will reveal entry points into their theological and ethical engagements with technology, which are covered in chapters 3–6.

Chapters 3 and 4 focus on Ellul and Virilio's multifaceted descriptions of technology's origins. It shows how, for Ellul and Virilio, the Industrial Revolution and its consequences reveal how modern technology came to be what it is today, but biblical texts like Genesis 3 and 4 reveal why humans ever produced technology. In other words, chapters 3 and 4 are where history and theology collide in Ellul and Virilio's work on technology. After the third chapter establishes the critical role *myth* plays in Ellul and Virilio's work on technology, the fourth chapter works backward, from the present, to the Industrial Revolution and its consequences, to Greco-Roman antiquity, to the Cain and Abel narrative in Genesis 4, to the Fall account in Genesis 3, all as they appear and clash with each other in Ellul and Virilio's work on technology. It moves in this direction to illustrate from sociological, historical, philosophical, and theological perspectives how and why Ellul and Virilio believe technology's potential to manipulate language and the senses produces modern forms of justification (Ellul) and mediation (Virilio) that are as damaging as they are questionable.

Chapter 5 shows how Ellul and Virilio expose the modern world's imprisonment within the nexus of technology, a net of material power and spiritual powers that was knit after creation's Fall as it is described in Genesis 3. If chapters 2, 3, and 4 focus on how Ellul and Virilio use sociology, history, philosophy, and theology to expose technology as an idol in the modern world, then chapter 5 presents the idol's schematics—what this idol does today and why it ought to be dismantled. The conclusion of this chapter also moves toward ethics as a precursor to chapter 6, in that it features some of the many ways Ellul and Virilio's work on material power and spiritual powers can help us identify, expose, and dismantle this and other idols.

The sixth and final chapter responds to the preceding chapters with ethics in and beyond Ellul and Virilio's work on technology. It describes the kinds of people we should be (2 Peter 3:11–18) in light of what Ellul and Virilio's work on technology reveals. As this chapter will show, there is much to be derived from Ellul and Virilio's ethics, but their ethics are more individually than communally focused. For this reason, chapter 6 both engages with Ellul and Virilio's ethics and develops complementary ethics to move their work into communal, especially ecclesial, territory. Then, it concludes with thoughts on how the nascent theological tradition featured throughout this book could be raised into maturity within the theological academy and the local church.

Having claimed that Ellul and Virilio's work on technology constitutes a nascent theological tradition, and having outlined the chapters that will support this claim, I will now frame the overarching thesis of this book with a final claim. Ellul and Virilio directly teach us how to identify, expose, and dismantle the modern world's idolatry of technology; they indirectly teach us how to identify, expose, and dismantle other forms of idolatry in the modern world that exist in the present or may exist in the future. The value of this teaching cannot be underestimated. As scripture reveals, history records, and scholars like Ellul and Virilio demonstrate, the idols that hold the most insidious power in the world are the idols that most people, including the people of God, do not think are idols. Yet, if we read Ellul's sociological work on technology as a sociology-history of technological idolatry; if we read Virilio's work on technology as a philosophy-phenomenology of technological idolatry; if we seek to understand how Ellul and Virilio use theology in their work on technology; and finally, if we engage with the conversation taking place in Ellul and Virilio's work, then we will be empowered as they were to identify, expose, and dismantle insidious idolatry in the modern world, whatever form it assumes.

NOTE

1. Some scholars have noted Ellul's influence on Virilio but none have examined this influence in detail nor investigated whether Ellul engaged with Virilio's work. Jacob Van Vleet, *Dialectical Theology and Jacques Ellul* (Minneapolis: Fortress Press, 2014), 101–102 is the most insightful treatment of the topic because it points out that Ellul influenced Virilio, identifies connections between their similar analyses of technological accidents and speed, and provides references to back up his claims. Douglas Kellner, "Virilio, War, and Technology," *From Modernism to Hypermodernism and Beyond*, ed. John Armitage (London: Sage, 2000), 118 and

122–123 cites Armitage's interview with Virilio and mentions general similarities between Ellul's, Heidegger's, and Virilio's analyses of technology. Finally, Adoni Alonso, "An Unseasonable Thinker: How Ellul Engages Cybercultural Criticism," in *Jacques Ellul and the Technological Society in the 21st Century*, from *Philosophy of Engineering and Technology* 13 (2013), eds. Jerónimo, Garcia, and Mitcham, 115–116 mentions in passing Ellul's influence on thinkers like Virilio and Jean Baudrillard.

Chapter 1

War, Modernity, and Technology

As the seed of each tradition grows in a particular type of soil, this chapter surveys the landscape of postwar France. It uses this survey to answer the following questions: Why did Ellul and Virilio convert to Christianity and how did their faith shape their lives and work? How did Ellul, an intellectual outsider from Bordeaux, connect with Virilio, an intellectual insider from Paris, and vice versa? Why did they write so much about technology and why did they write about it so critically? As this chapter will show, one word connects the answers to each of these questions: *war*, the theater in which the tragedy of technology and modernity is performed.

BRILLIANCE AND BRUTALITY

France, particularly Paris, produced some of the most important social, political, and philosophical thought of the twentieth century from the wreckage of war, shaping the thinking of Ellul and Virilio in pivotal ways. As postwar Europe struggled to rebuild, scholars and artists flocked to Paris, forming a unique intellectual culture. Historian Tony Judt describes this postwar milieu:

> Despite France's shattering defeat in 1940, its humiliating subjugation under four years of German occupation, the moral ambiguity (and worse) of Marshall Pétain's Vichy regime, and the country's embarrassing subordination to the US and Britain in the international diplomacy of the postwar years, French culture became once again the center of international attention: French intellectuals acquired a special international significance as spokesmen for the age, and the tenor of French political arguments epitomized the ideological rent in the world at large. Once more—and for the last time—Paris was the capital of Europe.[1]

But thirty years of violence made this brilliant culture brutal:

> Sartre and his generation claimed to have learnt from the war [the] inevitability of—and thus in certain measure the desirability—of political violence. This was far from being a distinctively French interpretation of recent experience: by 1945 many Europeans had lived through three decades of military and political violence. Young people all across the continent were inured to a level of public brutality, in words and actions, that would have shocked their nineteenth-century forbears. And modern political rhetoric offered a "dialectic" with which to domesticate calls for violence and conflict: Emmanuel Mounier, editor of the magazine *Esprit* and an influential presence on the Christian Left, undoubtedly spoke for many in 1949 when he asserted that it was hypocrisy to oppose violence or class-struggle when "white violence" was practiced on the victims of capitalism every day.[2]

Most scholars in this climate selectively critiqued the merciless fascism, communism, and capitalism of major world powers. Many were collaborators. Nicolas Nabokov, a defected Russian composer who worked with an organization of intellectuals and artists known as the Congress for Cultural Freedom (which was covertly funded and directed by the Central Intelligence Agency), is an example of both types of scholars. He once wrote in a letter to a colleague: "If successful [our project] will help to destroy the pernicious European myth (successfully cultivated by the Stalinists) of American cultural inferiority."[3] The project "will be a challenge of the culture of the free world to the un-culture of the totalitarian world and a source of courage and 'redressement moral,' in particular for the French intellectuals."[4] Ideologies and propaganda projects such as these were so common in postwar Europe that it transformed France into an arena of psychological warfare between Eastern and Western powers. But Ellul and Virilio avoided this tribalism. Instead, they resisted the violence of fascism, communism, and capitalism and they called out the duplicity of each intellectual vanguard. Their experiences during the war, their education, and their faith formed within them a singular conviction: tell the truth about what is going on, even if nobody wants to hear it. For Ellul, the trauma of war, Protestantism, and an education in history and sociology precipitated this conviction. For Virilio, it was the trauma of war; Catholicism; and studies in art, architecture, and philosophy.

The war troubles began for Ellul when he was removed from his professorship at University of Strasbourg for vocalizing his opposition to Marshall Pétain and the Vichy Regime.[5] After advising students to remain at their studies so as not to be forced to enlist in the German army, a student reported his subversive comments. Ellul lost his job but could have been arrested if a sympathetic official had not helped him avoid arrest. Other members in his family were not as fortunate, however. Ellul's father was arrested after it was discovered he was a British citizen, and eventually, he died in prison. Then, when it came to light that Ellul's wife Yvette was a British citizen, Ellul's family was forced to flee to rural France to avoid her arrest.

Once in Martres, Ellul farmed to provide food for his family. He needed support from local farmers in the village, however, because he was more of an intellectual than an agriculturalist. He also provided legal advice to locals, worked on a degree in theology that he nearly completed, and secretly taught university courses by distance. Still, it was difficult for Ellul to take care of his family, so his mother-in-law provided financial assistance that helped the family make it through the war.[6]

Ellul also joined the French resistance while all of the above was happening. He forged papers for, hid, and transported Jews and other persecuted people to the unoccupied regions of France. His family also provided lodging for resistance members. And while Ellul advocates nonviolence in his work, he does mention in interviews that he would have participated in the resistance's armed conflicts if he would have been able to find any of the weapons that were so scarce at the time.[7]

Finally, and most pressingly, Ellul lost one of his children in 1947.[8] Ellul does not speak about Simon's death, neither in interviews nor in published writing, but he dedicated *Meaning of the City* to Simon and he wrote a currently unpublished collection of poetry titled "The Book of Simon," an obviously intimate account of the loss of his son. In a conversation with Ellul's grandson Jérôme, he told me Simon's death "completely changed" Ellul's theology.[9] While Jérôme did not elaborate on how Ellul's theology changed after Simon's death, and it seemed inappropriate to ask for more details, it is reasonable to read anything theological written by Ellul after 1947 as work that has been inflected by Simon's death.

It is also noteworthy that the example of the automobile appears through Ellul's work as a symbol of the technologies so idealized in the twentieth century. For example, in his last major work on technology, he writes, "The automobile is [the] most perfect symbol of the technical society, and in buying it we are really buying a nexus of social symbols. It combines utility and futility, evasion and fatality, being elsewhere and finally meeting death."[10] The agony of the automobile accident that killed his soon looms over his writings, and within a storm of traumatic events, Ellul developed a writing plan that determined the next fifty years of his work.[11] When the war ended, he returned to teaching and writing.[12] This time it was in Bordeaux.

Whereas Ellul was in his late twenties, teaching, and exiled to rural France when the war began, Virilio was eight and exiled to Nantes with his family. It was occupied by the Germans, and because of the abominable absurdities of war, repeatedly bombed by France's allies.[13] After the war he was conscripted into the French military as a gunner, where he eventually became a cartographic officer. He was stationed in the French Occupied Zones in Germany and was deployed to Algeria for six months during its conflicts. Experiencing these traumas of war at a young age, and serving in the military during these formative years, shaped Virilio to the extent that war became a prominent theme in his writings.[14] He often identifies as a war baby, who was fathered and mothered by war, and educated in the university of war.[15] Such descriptions may seem exaggerated to readers new to Virilio, but the stories he shares about the war show he is neither dramatizing nor romanticizing what he was exposed to as a child. For example, he recalls in an interview:

> I wasn't told about [the concentration camps]. As [in], it wasn't something "people" talked about. There were the men who were executed, the hostages, the prisoners—I knew about that, I lived through that. But "people" didn't know about the camps. We found out only at the end of the war. And, [it] was an unthinkable, unplaceable thing for a child. You wouldn't believe it. [Basically], I didn't believe my eyes, until I saw the first concentration camp survivors getting off the trains.[16]

He continues:

> [The concentration camps were an] absolute horror. And then there was also an absolute escalation: there was Hiroshima. The war wouldn't go away. We were happy. The war was over. I remember [hearing] on the radio [. . .] a nuclear explosion. I don't remember if it was Bikini Atoll. . . . At any rate, "it" came over as a crackling or, in other words, like "static"—but we understood nothing about it, except for having an unconscious sense that it was space that was being rent asunder. Those who'd experienced the bombings understood; those who'd seen cities disappear understood; the sky's torn apart and bodies disintegrate. The atom bomb was an apocalyptic vision; you couldn't be glad about it. Camus wrote immediately in the papers that it was a scandal, whereas others said, "It had to be done, it's good." We, I, who had lived through the war and the bombings, I repeat that we couldn't be glad on this occasion. It was too much. It was like the concentration camps. Auschwitz and Hiroshima are pendants to each other. They've organized death, and that's the supreme perversion of the human spirit.[17]

I will say more in a later section about how Ellul and Virilio's experiences during the war provide critical context for their work on technology. However, it should be pointed out here that anecdotes like these show both thinkers should be read as scholars struggling to find ways to think and write after the Holocaust, Hiroshima, and Nagasaki; as people who witnessed concentration camp survivors off-boarding trains; and as citizens of a country caught between vicious world powers willing to drop atomic bombs on cities filled with civilians. Ellul tends to write and speak about these realities in detached and calculated ways. Virilio tends to write and speak about them in personal and passionate ways. But both scholars capture the truth that France's experience of World War II was far more immediate and devastating than that of North America's. This has to be kept in mind when reading the work of Ellul and Virilio because, as this book will illustrate from various perspectives, their proximity to the concentration camp and the atomic bomb viscerally shaped their lives and their work. So much, in fact, that concentration camps and atomic bombs appear throughout their writing as realities to be reckoned with.

Faith and Education

Ellul converted to Christianity at eighteen and eventually joined the smaller Reformed church rather than the dominant Roman Catholic church. His education in law and history and the intellectual influences of Søren Kierkegaard, Karl Marx, and Karl Barth were the primary reasons for this decision. He does not speak or write in detail about his conversion, but he does share some anecdotes in interviews. He grew up with a resolutely atheist father and a devout Christian mother who did not speak overtly about her faith but prayed fervently for her son's conversion. He was interested in the Bible at a young age, but his conversion was not prompted until a moving encounter with God in 1930. After this encounter, he briefly attended Catholic Churches and began reading to form his faith. He committed to the smaller Protestant community in France after reading Augustine and Calvin and feeling connected to their approaches to theology rather than those of Thomas Aquinas and Duns Scotus. Despite the early influence of Augustine and Calvin, it was Søren Kierkegaard and Karl Barth who had the most lasting impressions on his life and work.[18]

Ellul describes an early dialectal relationship forming in his thinking between a Marx-influenced sociology and history on the one hand, and a Kierkegaard and Barth influenced theology on the other, "[Marx] could explain to me my situation but not my human condition, my moral nature, my capacity to love, or my relationship with others."[19] But "the Bible and chapter eight of the Epistle to the Romans, had given me a response to an existential question." Thinkers like Augustine, Luther, and Calvin, and especially Kierkegaard and Barth, "[led] Ellul to an understanding of the Bible," which was "radical and disruptive."[20] He admired how these theologians stressed "there is no continuity between man's actions on earth and God's establishment of his kingdom. Or in another way, that one can't obtain salvation" through an "accumulation of works," "by a successful spiritual life," or "achieve good on [one's] own."[21] In other words, the Reformed church was a good fit for someone with Ellul's theological inclinations.

On the other side of Ellul's dialectic of theology and sociology was a formal education in law and an extensive knowledge of history. He once said in an interview, "I had always had a passion for history" and studying Roman law "plunged me back into Antiquity, into the atmosphere of Rome that I'd always liked [. . .] . I completely fell for the logical rigor of the construction of law."[22] In particular, "The method used by the Roman jurisconsults allowed for a constant adaption of the law to fit the politico-economic situation of the day."[23] From "a historians point of view I found the possibility of being able to change [the] things that didn't work well in the law absolutely prodigious."[24] Taken together, Ellul's esteem for Paul's Letter to the Romans, Luther the reformer, and Calvin, the lawyer meshed perfectly with his studies in Roman law and institutions. Once he joined the Reformed church, he was ever reforming, theologically, and sociologically.[25]

In contrast to Ellul, who resided in Bordeaux and intentionally distanced himself from the Parisian intelligentsia, Virilio moved to Paris after his time served in the military. He thrived in its intellectual culture. He attended École des métiers d'art (School of Artistic Crafts) and specialized in stained-glass artistry. His training led to his first profession working with renowned painters, sculptors, and architects such as Georges Braque, Henri Matisse, Le Corbusier, and Georges Roualt. He also audited lectures with phenomenologist Maurice Merleau-Ponty and was schooled in the thought of Edmund Husserl and Martin Heidegger. Then, a career in architecture materialized when he met Claude Parent and they formed and group called *Architecture Principe*. At one point, Virilio and Parent designed and built a church in Nevers, France, commonly known as *The Bunker Church*, which was dedicated to Saint Bernadette (discussed in chapter 6). Following participation in France's 1968 uprisings, he became a professor at École Spéciale d'Architecture at the request of students. He also formed collegial friendships with Jean Baudrillard, Gilles Deleuze, Félix Guattari, and Michel Foucault.

While Ellul and Virilio lived in different cities and were part of different intellectual and activists groups, their lives shared similar trajectories as they blended their faith and education in similar ways. In fact, Virilio also converted to Christianity at eighteen. According to him, one had to be either a communist or a Catholic after the war. He committed to Catholicism after interactions with French worker-priests, the left-leaning Catholics who lived and worked in lower-class neighborhoods after the war. Just as it did for Ellul, communism's brutality at the time inclined him toward Christianity. Still, both thinkers were influenced by the Marxist theory that formed so much of their landscape's thought and practice.[26] As for Virilio's theology, it has a more subtle place in his writing than an explicit presence. He shares reasons for this subtlety and speaks about Ellul's influence in the following interview:

> *John Armitage*: What role does [Christianity] play in your work? Do you see yourself, for example, as part of a French Catholic moral tradition that might include other Christian and existentialist critics of technology like Gabriel Marcel or Jacques Ellul?
>
> *Paul Virilio*: Yes, I do see linkages, especially with Jacques Ellul, rather than with Gabriel Marcel, who is from an earlier generation. But I cannot really place myself within what you call a Catholic tradition. The reason is that I have always been utterly unable to write about my faith. I do not have the gift for that. I have always considered that my life as a follower of Christ was something happening through my everyday life, not through my theoretical writings. It is not that I refuse to do it, I would gladly write a book about it, but I simply do not have the gift for it. You see, I do not have much of a theological culture. My conversion was an affair of the heart, a love affair you may say, more than an intellectual one. Speaking of religion, I feel much more at ease with an ordinary, poor person. When I am writing, I am somewhere else.
>
> *John Armitage*: In the late 1950s and throughout most of the 1960s the philosophy of structuralism began to challenge Christian existentialism, phenomenology and humanist Marxism. Structuralism was, of course, profoundly anti-humanist. Could your own theoretical approach be described as anti-humanist?
>
> *Paul Virilio*: Oh, not at all. I am an anarcho-Christian. It sounds quite paradoxical, but to me the definition of man is subsumed, and I quote it often, in a saying by someone I have come to like very much, Hildegarde from Bingen. St Hildegarde wrote, composed music, played the harp, and was many other things at once. The saying is: "*Homo est clausura mirabilium dei*": "Man is the closing point of the marvels of the universe" (i.e., God). Thus, for me, Man is not the centre of the universe, he is the end of the universe, the end of the world.[27]

It is noteworthy that Virilio affirms especial connections between he and Ellul, who was Protestant, rather than he and Marcel, who was Catholic. Certainly, the generational difference Virilio mentions is a factor, but similar to Ellul, Virilio describes himself as one who is both in and outside of his faith tradition. He identifies as a Christian, but he also expresses a felt distance from conventional forms of Christianity. In fact, he identifies as an anarcho-Christian, just as Ellul, the author of *Anarchy and Christianity*, did. Of course, his remarks about being "utterly" unable to write about his faith could seem to contradict my claim that theology influences his work.

However, he is being hyperbolic if not and evasive here. Even within the passage quoted he follows up comments regarding his supposed inability to speak or write about theology with a quote from Saint Hildegard that he skillfully applies to life and scholarship in the modern world. Setting aside this point for discussion until later chapters, though, Virilio's comment that he did "not have much of a theological culture" is more indicative of the reasons for his illusive response.[28] It can be interpreted in two different, but related ways, and both ways explain why Virilio and Ellul tend to be guarded when they discuss theology's presence in their life and work.

First, the comment can be taken to be mean that he either did not have access to or did not desire to pursue a theological education in the way Ellul did. This, of course, would have determined his ability or lack thereof to speak and write about faith and theology. Second, Virilio's remark can be interpreted as a reference to the intellectual culture he was most connected to, the Parisian intelligentsia. It was a crowd that did not tend to welcome anyone who identified as a Christian, even if they were unconventional ones like Virilio and Ellul. As Virilio once said in an interview, "I am a Christian. That is to say, I don't believe in death. And [Jean] Baudrillard didn't believe in life, that is the reality of life [. . .]. We were both conscientious objectors. Both atheists, but not the same kind: he didn't believe in reality, in particular in its acceleration, and I don't believe in death, that is to say, in cessation."[29] In another interview, he says this creates a tension in his writing: "I'm faced here with a problem of writing because I can't hide the fact that I'm a Christian (I don't see why I should hide it, for me it's essential), but on the other hand I don't want to use this 'advantage' to challenge things that are common to believers and non-believers alike."[30] Later, he expresses frustration with this tension: "We should be able to talk about [idolatry], but we can't. It's not politically correct. I can talk about it with Christians, but not with anyone else. And I'm fed up with that. We should be able to talk about sex, angels, God, everything. The world has arrived at an extremity where things need to be sorted out."[31] For these reasons, Virilio's lack of a theological culture reduced his ability, and in some cases his willingness, to feature theology in his work. But still, it is there.

Two anecdotes capture how unforgiving and threatening Ellul and Virilio's intellectual environment was. At one point, Ellul established friendly connection with Guy Debord after reading *Society of the Spectacle*. He eventually asked Debord if he could join the Situationists, a collection of intellectuals, avant-garde artists, and political theorists, who, along with various socially and politically charged acts which were more or less effective, played a significant role in the 1968 uprisings in France. But Ellul was denied membership because of his Christian faith. As for Virilio, he was threatened with physical violence and berated on the radio by artists and art critics because of what he wrote in *Art and Fear*.[32] Moreover, on top of possible social rejection and physical violence, there were career opportunities to be considered. Academic posts and publishing opportunities were scarce after the war, and scholars like Ellul and Virilio had to be strategic if they were going to secure let alone maintain teaching posts and publishing contracts. On the other hand, if constriction facilitates creativity, then the constrictions Ellul and Virilio felt because of their faith did have their positive effects. Over time, these constrictions inspired them to create unique ways to do theology: blend it with sociology, history, philosophy, and phenomenology to engage with and transform their intellectual landscape.

Strategic Style

In the most generic terms, Ellul and Virilio's approaches to writing about technology can be described as interdisciplinary. In more elaborate terms, their methodologies can be described as follows: for Ellul, dialectic simultaneously separates and integrates his sociological-historical and theological-ethical writing; for Virilio, fragment, in the style of the German Romantics, fuses theology with his philosophical-phenomenological writing.

Beginning with the general character of Ellul's dialectic, the sociological-historical side forms a Marxist view of society that cuts through surface social phenomena to analyze society's determinative structures.[33] An analogy Ellul uses to describe this method is an image of the ocean's strata.[34] On the ocean's surface are its waves: society's constantly changing phenomena. In the ocean's deepest parts are the dark depths, which power the middle to upper layers: society's metaphysical layer. In the ocean's middle is the tide, which is stirred by the ocean's depths and moves the waves on the ocean's surface: society's structural layer. Ellul said he primarily focused on the ocean's (society's) middle layer because it balanced realism and criticism without falling into the metaphysical detachments of a third layer analysis or the progressivist wanderings of a first layer analysis. In service to his structural method, he removes ornament from his prose, and he rarely, if ever, makes arguments, which cannot be verified by sociological and/or historical examples. From a theological perspective, Ephesians 4:14–16 offers a biblical frame for Ellul's use of ocean, waves, and wind imagery in these descriptions of his sociological method.

The theological-ethical side of Ellul's dialectic combines biblical theology, existential commentary, confrontation between the world's realities and revelations' truth, and ethics as a practical consequence of the preceding. As Ellul describes this method in *Ethics of Freedom*, "Hermeneutical freedom is [the] work to which we are committed to the utmost of our power in order that there may be the unveiling of sense which can come only by the act of the Holy Spirit."[35] Concretely speaking, this involves a three-step analytical process. First, "Appropriation; the message of God uses the cultural modes of a specific time and place and enters into them."[36] Second, "Contradiction; the difference between content of revelation and that of the particular culture becomes apparent."[37] Third, "Expropriation; the cultural schema or concept is absorbed by the content of revelation and the cultural sense is expropriated in favor of the revealed sense."[38] The influences of Kierkegaard and Barth are evident here, helpfully framing the general way in which Ellul's sociological and theological writings interact. There are contingent social structures into which the message of God appears (the sociological pole). There is a confrontation between contingent social structures and revelation's content (theological pole). Then, the contingent social structure (sociological pole) is seized and transformed by revelation's content (theological pole).

Ellul's dialectic was both a method in his writing and a structure for his writing, however. The majority of his books focus on either the sociological-historical pole or theological-ethics pole and there are only few books that combine both sides of this dialectic. For this reason, the primary question that preoccupies Ellul scholars is: To what extent should Ellul's sociological and theological texts be read conversantly? Near the end of his life, Ellul addresses the question. But similar to Virilio in the interview quoted earlier, he is at once open and closed about theology's place in his work:

Patrick Troude-Chastenet: In 1981 you told me that you had not in fact written several books but only *one* book, in which each separate "book" corresponded to one chapter. Does this mean *one* book for each register or compartment or that there is one book for the whole of your work; which brings us back to the question of the interconnection of your sociological works and your theological writings.

Jacques Ellul: The interplay [between] the two is very important for me as I have always thought that there should be a dialectical relation between a sociological and theological writing. It is not possible to read one without the other.

Patrick Troude-Chastenet: Since you have always kept the two distinct and always refused to see them fused into a single approach—in a closed circuit thought process—to now declare that all of your writing forms *one* single book surely undermines your wish to keep them separate.

Jacques Ellul: (Laughs) No, because they do belong to two different registers. They are inter-related if you like. In much the same way as a negative pole and a positive pole interact and then sparks fly between them. (Laughs)

Patrick Troude-Chastenet: Most sociologists and political scientists consider that your analysis of the technological society was determined by your religious convictions.

> *Jacques Ellul:* That is equally valid the other way round. I studied sociology as I studied history. I was trained in a rigorous discipline which required a lot of reading and a very critical approach to texts. It was in fact by studying the history of Roman law that I realised that our interpretation of the Bible is generally very insufficient and very superficial.[39]

How to harmonize without fusing Ellul's two registers? How to make the dialectical sparks fly in one's reading of Ellul? Does Ellul mean *poles* here in the sense of charged *batteries* that do not connect or charged *magnets* that pull together? The way I answer these questions here will focus predominantly on technology in Ellul's dialectical writings because this book focuses on technology in his work. To the first two questions, I suggest: reading Ellul's dialectical writings on technology through the hermeneutic of *myth* is a viable approach to answering these questions. For Ellul, *myth* is both the mysterious zone where the social meets the sacred and the explosive zone where revelation meets history. To the third question, I answer: the two dialectical *poles* are opposite *magnets* that attract each other in Ellul's writing. Although chapters 3 and 4 will show why I read Ellul in this way, it suffices to note here that he mentions in this interview, just as he did in the interview quoted earlier, that his understanding of Roman law shaped his reading of the Bible (see the previous section). This comment signals an important sociological, historical, and theological meeting point for his dialectical writings. For Ellul, the East, East of Eden, is the mythical cradle of urban civilization and the cradle of all technical action. The West, particularly Greco-Roman Antiquity, is the mythical ground zero for modern *technique*.[40] The former represents theological *myth* (revelation) as Ellul understands it and the latter represents sociological *myth* (history) as Ellul understands it. As these two poles (thesis/antithesis) in Ellul's dialectical writings on technology are drawn together like magnets, the sparks begin to fly.

Little has been written about theology's place in Virilio's philosophical-phenomenological writing.[41] As suggested in the previous section, part of the reason for this lack of engagement can be attributed to the types of scholars who tend to read Virilio's work. As Jean-Louis Violeau writes in the introduction to *Lost Dimension*, "Crypts and resurrection, ritual, call, vocation. There is an aspect of the person and the thought of Paul Virilio that is rarely underlined because it is out of fashion: his belief as a Christian and a Catholic."[42] Still, "[Virilio's] origins can be found there and so can some of the distant references in his thought: Babel and Noah's Ark, the Flood, the Apocalypse, the temptation in the desert, incarnation, total vision and the eye of God."[43] Whether or not theology is out of fashion today, Violeau's observation is basically true. Most Virilio scholars are not drawn to the theology present in his writing. He tends to attract readers interested in politics, art, architecture, and technology. But can Virilio's work really be appreciated if aspects of it are ignored? The same question could be asked of Ellul's work, which tends to attract scholars who are interested either in his sociology or theology.

Whether the question concerns Virilio or Ellul, my short answer is no. To really appreciate and understand Ellul and Virilio's work, one has to engage all of it. My extended answer to this question takes shape in this book as a whole, but I will add the caveat here that Ellul and Virilio adopted the writing styles they did so that much could be gained from reading their work even if aspects of it are set aside. A look at Virilio's use of the fragment style, and a quick look at Ellul's use of dialectical contradiction, shows how this is possible. Virilio does not use the word *fragment* to describe his writing style, but the frequency with which he quotes the German Romantics indicates familiarity with what Azade Seyhan describes as follows:

> Disillusioned by the political turmoil and what they saw as the crass materialism of [their age], [Friedrich Schlegel and Novalis] revive various forms and images of the exotic as alternatives to Western culture. They reimagine ancient Indian religions and medieval Christianity as cultures that endowed societies with meaning and integrity. They appreciate liminal and incomplete forms, tropes and histories as opposed to figures of arrest and closure and resist classical ideals of completion. In their elliptical and fragmentary [texts] they point to possibilities of self-revision and the generative power of poetic consciousness that transcends the structural and thematic limits of any given work of art. The early Romantics valued alliances of misalliances. [Novalis] contended that genuine philosophical thought called for the introduction of systemlessness. (*Systemlosigkeit*) into a system. Only such an interpretation could address the flaws of the system while avoiding the anarchy of systemlessness. The form that best captured a systemless system would be the Fragment, a form complete in itself but connected to the larger world like a microcosm.[44]

Virilio certainly laments his age's political turmoil and crass materialism. He also espouses affection for medieval Christianity and dabbles in eastern philosophy, history, and spirituality. Finally, and most pertinent to the current discussion, Virilio's writing is liminal, incomplete, systemless, elliptical, trope-laden, and micro and macrocosmic in scope. As Virilio puts it, "I work in staircases[.] I begin a sentence, I work out an idea, and when I consider it suggestive enough, I jump a step to another idea without bothering with the development. Developments are the episodes. I try to reach the tendency."[45] Or, using a different image, he says "I'm a painter who writes, you know! [My] books are very visual—they're very, very visual books! They're not words, they're visions!"[46] Which is to say, when Virilio forges fragments, builds incomplete staircases, and paints visions with words, he provokes readers to fuse the fragments, add the final steps, and interpret the visions in order to complete the work he has started in the text. He does not transmit meaning univocally in his writing, then, so much as he pushes the reader to form meaning as they engage with the text. As disorienting and frustrating this can be for new readers, it rewards the ones who persist with the work. Truth and beauty often emerge from the fragments of Virilio's writing if readers labor to fuse the pieces together.

It is also possible that Walter Benjamin's "constellation" approach to ideas and objects is an influence on Virilio's fragmentary style.[47] Similar to Benjamin, when Virilio has past, present, and future objects in view, he often attaches multiple perspectives, rather than a singular, synthesized perspective to the one object. Accordingly, if for Benjamin, "ideas are to things as constellations are to stars,"[48] for Virilio, the "stars" attached to the "constellations" of modernity so often are stars that speak of war and rumors of war, even if truth and beauty can appear among the war and the rumors of modernity. For this reason, as strange as it may sound, Virilio actually wants his readers to be disoriented and frustrated when they read his books. The phenomenological fragments of text he creates are intended to reflect the disorienting and frustrating state of the modern world; a crassly materialistic world constantly in political turmoil; a liminal, incomplete, systemless, elliptical, and trope-laden world. So there is an intentionality to the discomfort caused by Virilio's fragment style because a reader who is comfortable in this disorienting and frustrating world is less likely to search for truth and beauty than an uncomfortable reader. Virilio hopes for the second kind of reader, and if they are not that kind of reader when they come to the text, he will do his best to make them one. This makes Virilio quite similar to Ellul, who also aims to provoke readers, even if their methods differ. Whereas Virilio uses fragmented phenomenology to provoke, Ellul uses dialectical contradiction. He masterfully, and exhaustively, develops points and counterpoints and often uses irony to make his reader react. Whichever method they use, Ellul and Virilio's aims are identical: to snap readers out of lethargy so they will search for truth and beauty in a world which tends to obstruct such searches.

Taking these writing styles into consideration, it becomes clear why even the most committed readers of Ellul and Virilio gain such diverse insights from their work. If a reader is looking for insight into politics, they will find it in Ellul and Virilio's writing. If a reader is looking for insight into art, they will find it Ellul and Virilio's writing. If a reader is looking for insight into technology, they will find it Ellul and Virilio. And so on. My aim in this book is to help readers gain insight into theology and ethics in their writings on technology because these insights do not receive as much attention as others. To continue with this work, then, only two questions remain in this chapter. First, how did Ellul and Virilio connect with each other if they lived in different cities and were part of different intellectual and activist groups? Second, why did Ellul and Virilio write so much about technology and write about it so critically? The next two sections will answer these questions.

Connections

Given Virilio was the younger scholar, Ellul's influence on Virilio is more apparent than the reverse. Virilio openly states in interviews that he sees linkages between their work and he engages with Ellul at critical points in his writing.[49] But Ellul does cite Virilio once, in *The Technological Bluff*, which was published in 1988. The mention appears near the end of Ellul's writing career, just as Virilio's was gaining momentum in France. This citation and engagement implies that, had Ellul lived longer and continued to publish, more engagement likely would have materialized between the two scholars. Furthermore, correspondence by mail between Pierre Duriez and Ellul demonstrates that Ellul was aware of and engaged with Virilio's work.

When Ellul cites Virilio in *Technological Bluff*, he does so by quoting from an interview with Virilio that appeared in a special issue of *Le Monde* titled "Automobile et vitesse" (The Automobile and Speed) published in 1986. The exchange Ellul references features critical aspects of Virilio's early thought, lists some of Virilio's seminal publications, and captures in Virilio's own words the foundational themes that characterize his work from the 1970s onward. The interview in question is titled "Violence, Ralentir," (Violence, Slowdown) and Ellul quotes the opening paragraph to critique the automobile as a symbol of the twentieth century's powerful technologies and devotion to gadgetry.[50] His citation reads, "Speed is active violence. Faster, faster, but where to, and why?"[51] Since Virilio is known most for his thinking and writing on speed, this sentence represents the trajectory of his next forty years of writing, and by extension, Ellul's appreciation for and approval of his work.[52] Virilio seems to have reciprocated Ellul's appreciation by carrying forward what Ellul wrote in *Presence in The Modern World.* There, Ellul observes "we congratulate ourselves each time an airplane surpasses a speed record, and we make great efforts to succeed in going faster, as if speed were a good and sufficient goal in itself. But what is the point of saving time?"[53] In "this frightful round of unchecked means, nobody knows any longer where they are going, purposes are forgotten, and ends are overtaken. Human beings have set off at astronomically high speeds toward nowhere."[54] Thus, Ellul and Virilio join in asking the modern world: faster, faster, but where to and why? They also answer together: nobody seems to know, as they drive and fly at top speed toward nowhere.

It may also be that Virilio's analyses of speed and the automobile in this quote, and the entire interview, would have struck Ellul on an existential level. As mentioned earlier, Ellul lost his seven-year-old son Simon to the speed of an automobile thirty-eight years before this particular interview was published in *Le Monde*. Which is to say, Ellul's citation of Virilio in *Technological Bluff* marks a moment when the two scholars' lives and writings collide in one compact sentence. Yet, in what constitutes a curious and unfortunate blind spot, neither the French edition of *Le Bluff Technologique* nor the English edition of *Technological Bluff* account for this textual overlap between Ellul and Virilio and fail to provide a gloss that references the source interview with Virilio. Furthermore, the English translation does not even put the reference to Virilio in quotation marks. These are problematic oversights for both editions, but the reference nonetheless remains in Ellul's writing as a definite textual and historical connection to Virilio's work and thought.

Other hints in *Technological Bluff* suggest Ellul encountered Virilio's writing via other scholars' work. Namely, Jean Chesneaux's *De la modernité* (the English edition is titled *Modernity*) references Virilio's theory of *dromocracy* (also in relation to an analysis of the automobile), and Dominique Janicaud's *La puissance de la rational* (*Powers of the Rational*), references Virilio's theorization of the *war of time*.[55] Ellul states at the beginning of *Technological Bluff* that both books altered the form of this last major sociological work on technology.[56] It would be arguing from silence to claim the citations of Virilio's work in either text were an influence on Ellul's alterations to *Technological Bluff*, but it does imply that Ellul encountered Virilio as he was gaining prominence in the 1970s and 1980s.

The second explicit textual and biographical link that shows Ellul was aware of and engaged with Virilio's work appears in 1992, four years after the publication of the French edition of *Technological Bluff*, and two years before Ellul died in 1994.[57] It appears in a collection of correspondence between French sociologist Pierre Duriez and Ellul. The first letter in the collection shows Duriez writing to Ellul to inform him about Paul Virilio, a scholar he thinks "should interest" Ellul.[58] Apparently Duriez was unaware that Ellul cited Virilio in *Technological Bluff four* years earlier. Then, quite comically, Duriez writes, "To speak sincerely, the books of Paul Virilio annoy me and I cannot read them, because it is too brilliant, too 'elite university,' in short a little pedantic! But his ideas and thoughts are exciting and when he 'speaks,' I find it more understandable than when he writes for the intelligentsia!!!!"[59] Duriez includes the copy of the letter he sent to Virilio. It opens as follows: "I was 'struck' by your interview with Mark Hunyadi, in the *Journal de Genève* [on] the theme of speed and mutations that operate in the field of information."[60] Essentially, Durriez was "struck" enough by his encounter with the interview to think of Ellul and contact him. Perhaps this was because he saw an affinity between Ellul and Virilio and felt it was important to contact Ellul, providing not only the documents just mentioned, but also a draft and published copy of the interview with Virilio he mentions.

What is most revealing in this collection of documents is found in the published copy of the interview titled "Vitesse, pouvoir absolu [Speed, absolute power]."[61] The facsimile of the interview features significant annotations with yellow highlighter and black pen, all identifiably Ellul's, given the way he fastidiously collected and annotated books, articles, and newspaper clippings.[62] The most significant annotated passages from this interview are as follows, and they are worth listing in detail given how much each of them connect to the critical themes explored in this book:

- "The [Eastern and Western] arms race first led to the militarization of industry. This is the first militarization. The second is science."
- "Today, there is militarization of information, the advent of a military-information complex."
- "Communication weapons have gained supremacy. [Nuclear dissuasion] required the precision of the control of the military information. Now, whoever says military information says, in short, information."
- When Virilio speaks about a literature conference he attended: "There was a Bosnian author who, disgusted, deserted the federal army after four months [at the] front, then was taken prisoner by his compatriots before he managed to escape to France. 'Shame will survive me,' he says." Also, "There was [a] woman who read a page of her imaginary correspondence with her husband, cloistered in Sárajevo; the books drawn up in barricades in front of the windows are the best ramparts against the shells, he would write to her." Annotations begin at "Shame will survive me."
- Again when Virilio speaks about the literature conference: "A metaphor of what these writers testified in their flesh: in this desert, in this delirium, write protected. 'Write, write, write,' they said to themselves, because only literature will save us. The books are impermeable to shells, says the letter. Literature, 'a suicidal attempt to remain normal,' says one of the women writers, the same who will find, bitter: 'The only ones who will benefit from this war are the artists.' " Specifically "Literature, 'a suicidal attempt . . .' " is annotated.
- Finally: "I want to say: I pass the hand to those who can use my work to find solutions. I am a watchman who spots the tendencies and temptations of our situation. Democracy is threatened by technology; not by machines, against which I have nothing, but by the speed implemented by machines. This threat would require a collective intelligence of technique, it is the first parade. If the screen is the last frontier, you have to

go back to writing. Dialogue is born democracy, the social bond. Writing conceals an irreplaceable communication property. Facing the screen, the written."[63]

As the chapters to follow will demonstrate in detail, the militarization of technology and science in the aftermath of World War II; the ramping up of the Cold War; the pervasive and dangerous power of manipulated discourse and information; the critical force and diagnostic potential of literature; images of the delirium in the desert; and the biblical image of the watchman all play critical roles in the Ellul and Virilio's work on technology. These themes will receive detailed examination in due course. For now, it suffices to point out for a second time that there is not only a thematic, stylistic, and biographical affinity between Ellul and Virilio, but also verifiable historical and textual connections.[64] With this in view, the next section will explain why both scholars devoted so much of their writing to technology.

The Concentration Point

Ellul and Virilio's traumatic experiences during the war years alerted them to the dangers of technology, many of which were downplayed by the leaders of nations, academics, and the masses as Europe attempted to reconstruct after the war. On the one hand, technological innovation made it possible for World War I and II to reach scales of devastation that surpassed any other war in history. On the other hand, technological innovation made it possible for nations to recover in the postwar years, and eventually, helped raise the standard of living in Europe and inaugurate an era of mass consumerism that began in North America and spread around the world.[65] How could anyone object to the expansion of technology after the war if it appeared to promise a good life at a low cost after so much suffering? Ellul and Virilio could object because they did not want to see more suffering caused by power hungry nations pushing technological innovation in order to secure more power.

Some of the anecdotes examined in a previous section capture Ellul and Virilio's traumatizing encounters with the brutal technologies brought into existence during the twentieth century for the purposes of killing and controlling people. Virilio's story about learning of the concentration camps and the dropping of atomic bombs on Bikini Atoll, Hiroshima, and Nagasaki is an obvious example. Yet there are many other examples that capture the power and violence of technologies that were alarming to Ellul and Virilio even if they seem harmless to current readers. Focusing on the mass communication technologies that helped dictators like Mussolini and Hitler ascend to power, Virilio writes in *War and Cinema*:

> In Western Europe and the Soviet Union, a new breed of military and revolutionary leader was beginning to have [a] charismatic effect on the masses[.] Real power was now shared between the logistics of weaponry and of sound and images, between war cabinets and propaganda departments[.] "Propaganda is my best weapon," declared Mussolini. And this is how Rosavita described the Duce: "Go, Caesar . . . ! Your task is fulfilled. From Caesar has come Benito Mussolini, strong and powerful as History has never before displayed; *his will partakes of the supernatural, the Divine, the miraculous, of Christ among men.*" Just as unexpectedly, when Hitler was asked in private who was his great historical model, his answer was not Bismarck, as one might have thought, but Moses. Perhaps it has not been properly understood that these miracle-working dictators no longer ruled but were themselves *directors.*[66]

He continues, describing Hitler's use of the cinema and cinematic effects:

> Hitler, who closely observed the crowds flocking to celebrate the black masses of cinema, declared one day in 1938: "The masses need illusion—but not only in theaters or cinema. They've had all they can take of the serious things in life." The Nazi *Lebensraum* was less the fulfillment of Bismarck's grand political schemes—although these formed the substance of Hitler's political speeches—than the transformation of Europe into a cinema screen, for a people "suddenly horrified by the everyday, the ordinary, and fascinated by the unusual" (Leni Riefenstahl). Hitler violated only everyday realism, and the very nature of his crimes cannot be understood unless we remember his extraordinary technical knowledge of stage-direction, trick photography, trapdoor devices, revolving stages and, above all, the varied potential of illumination and floodlighting. "Hitler may not have been the great statesmen we saw in him," argues Albert Speer, "But he was and remains a psychologist whose like I have never encountered. Even as supreme army commander, he thought more about the psychological effect of a weapon than about its operational force." (This from Speer, whose achievements include the Stuka sirens and A4 missile warheads.)[67]

Leni Riefenstahl, a filmmaker, actress, and artist, directed the Nazi propaganda films *Triumph of the Will* and *Olympia.* Albert Speer, an architect, was Hitler's minister of Armaments and War Production. Both disingenuously denied knowledge of the Holocaust. Such individuals appear throughout Virilio's writing and they will reappear at later points in this book as examples of technicians who aided people in power to commit atrocious acts and abdicated responsibility for their own actions. Here, however, the cinema's power and violence as Virilio describes it require more attention than Riefenstahl and Speer's stories.

When most people look at a cinema today, all they see is a cinema. When Virilio and Ellul looked at a cinema, they saw war. Such reactions may seem exaggerated and alarmist to a current reader, but if this is the case, the following question is worth considering: What would have Mussolini or Hitler been without the power of mass communication? So much less than what they were *with* the power of mass communication. Radio, television, and the cinema may appear innocuous when they are compared to the communication technologies we have access to today, but Ellul and Virilio constantly remind readers that any communication technology can help a dictator seize power and mobilize a nation. So, another question may be asked at this point: If Mussolini said propaganda was his most effective weapon and Hitler was more concerned about the psychological effects of a weapon than its operational force, how concerned should we be about current communication technologies and their psychological effects?

Just as Virilio takes aim at mass communication technology's connections to war in *War and Cinema*, Ellul takes aim at mass consumer technology's connections to war in *Technological Society*. He writes:

> For a long time it was believed that technique would yield a harmonious society, a society in equilibrium, happy and without special problems. This society would resign itself to an easy life of production and consumption based on an untroubled commercial ideology. This model of bourgeois tranquility seemed to correspond exactly to the preoccupations of technology. The *summum bonum* was comfort[.] The sudden plunge of the technically most advanced societies was into war and mutual destruction was a rude awakening for the bourgeois. An aberration? Scarcely. It had been forgotten that technique means not comfort but power. The bourgeois countries had developed their technical systems at a comfortable pace, until these systems had fully exploited their possibilities of orderly growth. Then technology, with its accelerated tempo, took over. The smaller nations were unable to keep up. And the great technical countries had willy-nilly to abandon their languid pace and accommodate themselves to the real tempo of the technical society. The result was the disproportion between the leisurely bourgeois mentality and the explosive tempo of technique to which we give the name war. A by-product of this ecstasy was a certain mystique. The American myth was born, presenting exactly the same religious traits as the Nazi or communist myth.[68]

There are two sides to technological innovation. The first is the side most people, especially the bourgeois, like to focus on. Namely, the appealing things that technological innovation promise such as harmony, happiness, and comfort. The second is the side most people, except the ambitious, the perceptive, and the realists, do not like to focus on. Namely, what is required to attain the appealing things technological innovation promises. Whatever side is focused on, the first side's appeal drives innovation and the second side steers the innovation toward disaster.

Ellul uses the example of atomic energy elsewhere in *Technological Society* to expose this relationship. The research and design required for the military use of atomic energy is simple in comparison to that required for industrial use (it is easier to explode an atom than it is to use an atom to safely power a city). Consequently, when scientists wanted to research and design atomic power for industrial uses they had to research and design it for military uses. But once the atomic bomb was made, who or what was able to stop it from being used? Ellul answers: "In the interim period represented by the bomb, the possessor, finding himself with so powerful an instrument, is led to use it. Why? Because everything which is technique is necessarily used as soon as it is available, without distinction of good or evil. This is the principle law of our age."[69] Once something as powerful as the atom bomb was made it became too tempting not to use, even if it initially was considered—at least by the scientists leading the project—a potentially destructive phase that was a part of a broader research project for constructive purposes. For this reason, Ellul contends in this passage and many others that *war* simply is the word the modern world uses to describe the disparity between what is promised by technological innovation—comfort and happiness—and what is produced—power and destruction. It is this disparity—war—which drastically increases the speed at which the modern world moves through history.

Although Virilio does not reference Ellul in *Speed and Politics*, I think he had passages like these in mind when he wrote one of his most quoted axioms "history progresses at the speed of its weapon's systems."[70] Regardless, Ellul and Virilio agree that the urgency of war tends to bring about the kind of technological innovation required to move history from one phase to the next. History's transition into modernity certainly is a case in point. So many of the technologies that have shaped the twentieth and twenty-first centuries, including the computer and the internet, were generated by the urgency of war. So, if there is a metonym for *modernity* in Ellul and Virilio's work, it is *technology*, and both modernity and technology are what they are today because of the wars that happened at the start of the twentieth century. Note, I am not suggesting, just as Ellul and Virilio do not suggest, that modernity and technology necessarily are beholden to war. Rather, I am suggesting, along with Ellul and Virilio, that if we do not examine the connections between war, technology, and modernity, then it is more likely that modernity and technology will be beholden to war.

The final elements to note in these passages are the religious elements and the theological language. Virilio describes how communication technologies enabled the deification of Mussolini and Hitler. Ellul describes how consumer technologies inspired religious fervor for Nazi, communist, and capitalist nations. It may offend American readers that Ellul places America's propaganda at the same level of Nazi and communist propaganda, but Ellul is not alone in the equation. Virilio agrees. In fact, in the quote referenced earlier from *War and Cinema*, Virilio frames his descriptions of Hitler and Mussolini's use of the cinema with a description of Cecil B. De Mille, the famous founding father of American cinema. He writes, "It is said that by the time De Mille was finishing *The Ten Commandments*, he imagined he was God himself leading the Jewish people. His dictatorial attitude [suggested] a man possessed, and his assumption of divinity involved a kind of charismatic infallibility, stemming from foreknowledge of scripts."[71] For "a whole generation of cinematic miracle-workers, the process of direction, even if improvised, literally took the form of *revelation*—that is, divine action which makes known to men truths that they would not be able to discover by themselves."[72] Fascist, communist, and capitalist propaganda, along with the technologies used for its dissemination, were functionally identical from the perspective of someone living in postwar France. Ellul and Virilio saw firsthand how propaganda, regardless of the nation or leader it represented, could mobilize the masses for war as it apotheosized people and nations.[73] For this reason, the war alerted Ellul and Virilio to the emergence of an idol in the twentieth century: technology, the efficient apotheosizer and mobilizer, the concentration point of modernity's power. From their positions on the ground in postwar France, they resolved to warn anyone who would listen: watch out, idols promise power and demand blood. And so began their interdisciplinary work to identify, expose, and dismantle this idol. With this in view, the next chapter examines the sociological-historical and philosophical-phenomenological facets of this work.

NOTES

1. Tony Judt, *Postwar: A History of Europe since 1945* (London: Vintage Books, 2010), 210. For further context on postwar Europe see 13–129.

2. Ibid., 211. As a part of his involvement with the Personalist movements in France, Ellul knew and worked with Emmanuel Mounier and published in *Esprit*. Eventually, Ellul broke with Mounier and *Esprit* over theoretical and practical disagreements with over, many other things, approaches to political activism such as the ones described in this passage.

3. Frances Stonor Saunders, *The Cultural Cold War: The CIA and The World of Arts and Letters* (New York: The New Press, 1999), 95. In fact, Denis de Rougement—whom Ellul knew through work with *L'Ordre Nouveau*—was a part of the Congress for Cultural Freedom. See 85; 78–79; 89; 97–101; 187; and 329–333

for details. Given the covert nature of the Congress's relationship with the CIA it is unlikely Ellul knew of de Rougement's involvement with the CIA. But Ellul would surely have been irate had he discovered it.

4. Ibid.

5. There already are numerous works that address Ellul's biography so I will only mention biographical details that are pertinent to this chapter and book. See Jacques Ellul and Madeleine Garrigou-Lagrange, *In Season Out Of Season*, trans. Lani K. Niles (Cambridge: Harper and Row Publishers, 1981); Jacques Ellul, *Perspectives On Our Age*, ed. Willem H. Vanderburg, trans. Canadian Broadcasting Corporation (Toronto: House of Anansi Press, 2004); and Jacques Ellul and Patrick Troude-Chastenet, *Jacques Ellul on Politics, Technology, and Christianity*, trans. Joan Mendès France (Eugene: Wipf and Stock, 2005) for the best primary sources in English for Ellul's biography. Andrew Goddard, "Introduction to Jacques Ellul's Life and Work," *Living the Word, Resisting the World: The Life and Thought of Jacques Ellul* (Carlisle: Paternoster Press, 2002) is a secondary source, but it is the most comprehensive biography of Ellul currently published in English.

6. Ellul and Garrigou-Lagrange, *In Season*, 46–49.

7. Ellul and Troude-Chastenet, *Politics, Technology, and Christianity*, 77.

8. Jeffrey Greenman, Read Schuchardt, and Noah Toly, *Understanding Jacques Ellul* (Eugene: Cascade Books, 2012), 34. The book mentions that Simon was killed by a vehicle when he was crossing the street but does not provide further details.

9. Jérôme Ellul, interviewed by author, Vancouver, June 28, 2018.

10. Jacques Ellul, *The Technological Bluff*, trans. Geoffrey W. Bromiley (Grand Rapids: Eerdmans, 1990), 372.

11. Ellul mentions this plan in various books and interviews. An example is Jacques Ellul, *Presence in the Modern World*, trans. Lisa Richmond (Eugene: Cascade Books, 2016), xviii and Ellul and Garrigou-Lagrange, *In Season*, 70–71.

12. Ellul does not speak at length about his personal experiences during the war much beyond what is shared in this section. But he often uses the example of World War II from his early to his late writings. It is a telling but typically underemphasized indication that the war significantly impressed itself on his person and thought, especially on his studies of technology. See, for example, Jacques Ellul, *Autopsy of Revolution*, trans. Patricia Wolf (New York: Alfred A. Knopf, 1971), 28–29; Jacques Ellul, *The Ethics of Freedom*, trans. Geoffrey W. Bromiley (London: Mowbrays, 1976), 376–377; Jacques Ellul, *The New Demons*, trans. C. Edward Hopkin (New York: Seabury Press, 1975), 166–202; Jacques Ellul, *The Betrayal of the West*, trans. Matthew J. O'Connell (New York: Seabury Press, 1978), 127–128.

13. For the most current and concise summary of Virilio's biography, see Mark Lacy, "Paul Virilio (1932–2018)," *Verso*, September 20, 2018, https://www.versobooks.com/blogs/4042-paul-virilio-1932-2018, written just after Virilio's recent death.

14. Paul Virilio and Marianne Brausch, *A Winter's Journey*, trans. Chris Turner (London: Seagull Books, 1997), 3–64 and Paul Virilio and Bertrand Richard, *The Administration of Fear*, trans. Ames Hodges (Los Angeles: Semiotexte, 2007), 13–28.

15. See Paul Virilio, *L'insécurité du territoire* (Paris: Éditions Galilée, 1993), 11–21 for a hauntingly beautiful and harrowing phenomenological, but untranslated, description of Virilio's childhood during the war. For a translated description, see Paul Virilio and John Armitage, *Virilio Live: Selected Interviews* (London: SAGE Publications, 2001), 16 and 37. All translations from the French are my own unless otherwise noted.

16. Virilio and Braush, *A Winter's Journey*, 56.

17. Ibid., 58.

18. Ellul and Garrigou-Lagrange, *In Season*, 16–17 and Ellul and Troude-Chastenet, *Politics, Technology, and Christianity*, 49.

19. Ellul and Garrigou-Lagrange, *In Season*, 15.

20. Ibid., 59–60.

21. Ibid.

22. Ellul and Troude-Chastenet, *Politics, Technology, and Christianity*, 71.

23. Ibid.

24. Ibid.

25. Ellul and Garrigou-Lagrange, *In Season*, 84. On pages 85–97 and 108–116; Ellul describes his approach to the reformative work he did in the institutional church in partnership with his friend Jean Bosc. Although their research, planning, and initial implementation of new curriculum and programs, which was carried out over the course of several years, was rejected by the national council, it was a remarkable reformative effort led by Ellul and Bosc. For more academic examples of Ellul's understanding of the Reformation and Reformed theology and ethics, see Jacques Ellul, "The Contemporaneity of the Reformation," *Sources and Trajectories*, trans. Marva Dawn (Grand Rapids: Eerdmans, 1997), 135–165 for a translation of an early article published in 1959 in which Ellul writes about his understanding of the Reformation and how it relates to his historical moment. There also is an even earlier article that Ellul wrote in 1945, which similarly explores the historical importance of the Reformation. See Jacques Ellul, "Signification actuelle de la Reforme," *Protestantisme Francais*, ed. Marc Boegner and André Siegfried (Paris: Plon, 1945), 137–165. For an English translation of the article, see Jacques Ellul, "Current Signification of the Reformation," *European Journal of Theology* xxvi (2017), no. 2, trans. Jacob Rollison, 98–113.

26. Virilio, *Virilio Live*, 19–21.

27. Ibid., 20–21.

28. Ibid.

29. Paul Virilio, *Grey Ecology*, trans. Drew Burk (New York: Atropos Press, 2009), 42–43.

30. Paul Virilio and Sylvère Lotringer, *Pure War*, trans. Mark Polizzoti (Los Angeles: Semiotexte, 2008), 141.

31. Ibid., 203.

32. Julie Rose, "Translators Preface," *Art and Fear*, trans. Julie Rose (London: Continuum, 2000), vii.

33. Ellul and Garrigou-Lagrange, *In Season*, 27–28.

34. Jacques Ellul, *Hope in Time of Abandonment*, trans. C. Edward Hopkin (New York: Seabury Press, 1973), 279–281.

35. Ellul, *Ethics of Freedom*, 163–164.

36. Ibid., 164.

37. Ibid.

38. Ibid.

39. Ellul and Troude-Chastenet, *Politics, Technology, and Christianity*, 90–91. For more on Ellul's dialectical method, see Jacques Ellul, "On Dialectic," *Jacques Ellul: Interpretive Essays*, eds. Clifford G. Christians and Jay M. Van Hook, trans. Geoffrey E. Bromiley (Chicago: University of Illinois Press, 1981), 291–308.

40. Jacques Ellul, *Meaning of the City*, trans. Denis Pardee (Eugene: Wipf and Stock, 2011), 14. Jacques Ellul, *The Technological Society*, trans. John Wilkinson (New York: Vintage Books, 1964), 27–28.

41. Jacques Arnould, *L'Apocalypse du vide* (Paris: EdiLivre, 2019); Neil Turnbull, "Light and Illumination: Paul Virilio's Neoplatonism and Theological Critique of Modernity," *Cultural Politics* 11, no. 2 (2015): 260–274 and Chris Huebner, "Patience, Witness, and the Scattered Body of Christ: Yoder and Virilio on Knowledge, Politics, and Speed," *A Precarious Peace* (Waterloo: Herald Press, 2006), 115–132; and Michael Hirsch, "Complexity in the Age of Simplicity," *Grey Ecology*, 115–131 are rare examples. I engage with these pieces in later chapters. For good general introductions to Virilio's writing, see Mark Lacy, *Security, Technology, and Global Politics: Thinking with Virilio* (London: Routledge, 2014) and Ian James, *Paul Virilio* (New York: Routledge, 2007).

42. Jean-Louis Violeau, "Introduction: Critical Space under Nuclear Winter," *Lost Dimension*, trans. Daniel Moshenberg (Los Angeles: Semiotexte 1984), 20.

43. Ibid.

44. Azade Seyhan, "What Is Romanticism, and Where Did It Come From?", *The Cambridge Companion to German Romanticism*, ed. Nicholas Saul (Cambridge: Cambridge University Press, 2009), 17–18. See Paul Virilio, *The Aesthetics of Disappearance*, trans. Philip Beitchman (Semiotexte, 1980), 87 for one of many passages in which Virilio quotes Novalis. In this case, he works with the notion of the "microcosm" as it is described in Seyhan's discussion of the German Romantic's use of fragment.

45. Virilio, *Pure War*, 52.

46. Virilio, *Virilio Live*, 160.

47. See Walter Benjamin, "Epistemo-Critical Foreword," *Origin of the German Trauerspiel*, trans. Howard Eiland (Cambridge: Harvard University Press), 1–40 for discussion of Benjamin's constellation. Like Virilio, Benjamin was engaged with the German Romantics, so it could be that Virilio was brought to the Romantics via Benjamin, particularly works like *The Concept of Art Criticism in German Romanticism*.

48. Ibid., 10.

49. Virilio, *Virilio Live*, 20–21. Paul Virilio and Philippe Petit, *Politics of the Very Worst*, trans. Michael Cavaliere (New York: Semiotexte, 1996), 95.

50. Paul Virilio and Frédéric Gaussen, "Violence, Ralentir," *Le Monde*, June, 1986.

51. Jacques Ellul, *Technological Bluff*, trans. Geoffrey W. Bromiley (Grand Rapids: Eerdmans, 1990), 266–267. Jacques Ellul, *Le Bluff Technologique* (Paris:

Pluriel, 2012), 284–285. Paul Virilio and Frédéric Gaussen, "Violence, Ralentir," *Le Monde*, June, 1986. These are the three texts that, to one degree or another, given the quality of translation and editing, show Ellul's citation of the interview with Virilio.

52. Although Virilio never cites it directly and takes his analysis of technology and speed into different territory than Ellul, it is possible chapter "The Problem of Acceleration" in Jacques Ellul, *The Technological System*, trans. Joachim Neugroschel (New York: Continuum, 1977), 283–309 influenced Virilio's work on speed and technology.

53. Ellul, *Presence in the Modern World*, 44–45.

54. Ibid.

55. Jean Chesnaux, *De la modernité* (Paris: Maspero, 1983), 16. Dominique Janicaud, *Powers of the Rational*, trans. Peg and Elizabeth Birmingham (Bloomington: Indiana University Press, 1994), 261. It is possible Virilio's thinking on the *war of time* and *chrono-politics* is influenced by Jacques Ellul, *Political Illusion*, trans. Konrad Kellen (New York: Vintage Books, 1967), 27–29 and 49–67. See Virilio and Lotringer, *Pure War*, 7–13; 51; 126–128; 140 for examples of Virilio's thinking on the *war of time* and *chrono-politics.*

56. Ellul, *Technological Bluff*, xiv.

57. Pierre Duriez to Jacques Ellul, December 1, 1992, Manuscrits des universités de Bordeaux, Manuscrits de Sciences Po Bordeaux, Carton 1/117, http://www.calames.abes.fr/pub/ms/Calames-20131101113749099.

58. Ibid.

59. Ibid., Author's emphasis.

60. Ibid.

61. Paul Virilio and Mark Hunyadi, "Vitesse, pouvoir absolu," *Journal de Genève et Gazette de Lausanne* no. 268 (November 14, 1992), 22.

62. I was intimately acquainted with Ellul's collection and annotation of books, articles, and newspaper clippings when I visited the Ellul family's personal archives in Mont de Marsan. Among the many treasures I discovered there, I encountered folders of Ellul's that were packed with research material. Usually these folders contained anything from academic journal articles, to newspaper clippings, to pages torn out of books, to messy scrawls on scrap pieces of paper. They were annotated in a consistent style, which resembles the annotations in the copy of the interview Ellul received from Duriez.

63. Ibid.

64. Other potential links between Ellul and Virilio I have identified are that: Ellul, in his later writings, cites Jean Baudrillard's work; that Virilio as early as 1969 published articles in *Esprit* (which Ellul also wrote for and read); and, Ellul engaged with Maurice Merleau-Ponty's and Edmund Husserl's work. Virilio was a friend of Baudrillard's and is featured at places in Baudrillard's work, implying Ellul could have encountered Virilio's name and possibly his writings therein. Likewise, Ellul engages with *Esprit* and *Le Monde* in his work and may have encountered Virilio's writing or name, other than the interview, which is quoted in *Technological Bluff*. Ellul also mentions Merleau-Ponty and Husserl in his work, and, as mentioned previously, Virilio studied under Merleau-Ponty, who was, in turn, a student of Husserl's. Ellul

also mentions Abbé Pierre, a French priest who founded the Emmaus Community, and Virilio worked with Abbé Pierre. Finally, Ellul mentions Henri Lefebvre, Yona Freidmann, and Le Corbusier, who Virilio either worked with or mentions being acquainted with during his partnership with Claude Parent in the *Architecture Principé* group. See *New Demons*, 70; 97; and 145; *Autopsy of Revolution*, 120–123 and 278; *Technological System*, 260–261 and 343; *Betrayal of The West*, 89; Jacques Ellul, *Empire of Non-Sense*, trans. Michael Johnson and David Lovekin (Berkshire: Papadakis, 2014), 113; and Jacques Ellul, *Freedom, Love, and Power*, trans. Willem Vanderburg (Toronto: University of Toronto Press, 2010), 74–75 for references.

65. See Victoria de Grazia, *Irresistible Empire: America's Advance through 20th-Century Europe* (London: Belknap Press, 2005) for a remarkable and comprehensive study of how, "In the process of challenging Europe's bourgeois commercial civilization and overturning its old regime, the United States established its legitimacy as the world's first regime of mass consumption" (5).

66. Paul Virilio, *War and Cinema*, trans. Patrick Camiller (London: Verso, 1984), 67–68.

67. Ibid. Ellul agrees with the assessment, though he does so in terms more sober than Virilio. For example, see Ellul, *Technological Society*, 83.

68. Ellul, *Technological Society*, 421–422.

69. Ibid., 99.

70. Paul Virilio, *Speed and Politics*, trans. Mark Polizzotti (Los Angeles: Semiotexte, 2006), 90.

71. Virilio, *War and Cinema*, 66.

72. Ibid., 66–67.

73. Jacques Ellul, *Propaganda*, trans. Konrad Kellen and Jean Lerner (New York: Vintage Books, 1975, first published: 1962), ix–xiv.

Chapter 2

The Construction of Idols

The previous chapter ended with the suggestion that Ellul and Virilio's work on technology can be read as the identification, exposure, and dismantling of technological idolatry. It is tempting to pass over the sociology, history, philosophy, and phenomenology in their work on technology and move to theology and ethics. The value of the former aspects of Ellul and Virilio's writing, however, is that it urges those of us who work within the disciplines of theology and ethics to avoid hurried analyses of idolatry. As this chapter will demonstrate, Ellul's sociology and history and Virilio's philosophy and phenomenology of technology help us answer questions that many theologians and ethicists do not answer adequately, if they bother to ask them at all: How is an idol constructed and what does an idol do? This chapter focuses on the essential concerns, terms, and methods in their work on technology to show how Ellul and Virilio help us answer these questions. For this reason, I do not feature an exhaustive study of Ellul's sociology and history of technology and Virilio's philosophy and phenomenology of technology in this chapter. To provide such a study would require an entire book, if not multiple books. Instead, this chapter focuses on Ellul and Virilio's reformulations of the classic Greek concepts *tékhnē*, *poiēsis*, and *epistēmē,* because they provide illuminating entry points into their extensive writings on technology. For Ellul, modern technology's mutation of this classic trio signals a threatening manipulation of language that produces a new form of justification; for Virilio, it signals a threatening manipulation of the senses that produces a new form of mediation; and, for both Ellul and Virilio, this mutation and production is the primary means by which the idol of technology was constructed in the twentieth century.

TECHNIQUE AS A RATIONAL PROCESS

Ellul uses the French terms *la technique*, *la technology*, and *la propagande* in ways that do not survive in standard English translations. Defining the terms *tékhnē*, *poiēsis*, and *epistēmē*, and establishing how they relate to *technique* as Ellul defines it, explains why he was so concerned by its power in the modern world. The trio's classic definitions are:

- *Tékhnē*, meaning craft, art, skill, or making
- *Poiēsis*, meaning practice, doing, or acting
- *Epistēmē*, meaning knowledge, science, or understanding

While this trio implicitly informed Ellul's understanding of *technique* as a rational process, there was another prominent scholar of technology who explicitly used the trio in his writings. Martin Heidegger, whose 1949 lectures on technology and publication of those lectures in 1954, draws on Plato and Aristotle to inform his critiques of technologies.[1] In contrast to Heidegger, Ellul's early works on *technique* in partnership with Bernard Charbonneau in the mid- to late-1930s, and work on *Technological Society* beginning in 1947, do not engage at length with Greek philosophy.[2] But Ellul read the Greeks and often critiques modern accounts of epistemology rooted in the Greek's philosophical accounts of *epistēmē*.[3] He also forms connections between *epistēmē* and *technique* and refers to *tékhnē* and *poiēsis* at various points in his writing to compare and contrast it with his descriptions of modern *technique*.[4] An example that especially signals the connection between classic *tékhnē* and modern *technique* is Ellul's allusion to Diderot's description of a painter's *technique* in *Technological System*.[5]

The classic question of *tékhnē* concerns the artist or craftsperson and his or her respective skills. In the case of a painter, how does the painter's ability, training, senses, thoughts, experiences, and environment determine their decision to use this or that instrument, this or that color, this or that movement of the brush, and so on? Such questions, and how they are answered, characterize the classic process of *tékhnē* that emerges from *epistēmē* as knowing and *poiēsis* as acting in the world. The way Ellul poses and answers such questions about *tékhnē* in its modern setting demonstrates how *tékhnē*, *poiēsis*, and *epistēmē* quantitatively and qualitatively mutate with the advent of the modern era.

For Ellul, the Industrial Revolution triggers a mutation of *tékhnē*, *poiēsis*, and *epistēmē* because it is the moment in history when the ideal of efficiency started to determine increasingly mechanizing economies and their societies. As the mechanization of industry exponentially increased the means of production's efficiency, the cost of labor and commodities exponentially decreased, thereby creating a cycle of production and consumption driven by demands for ever-more efficient means to keep the cycle moving. Basically, once a society becomes used to paying less money for more commodities, its economy has to keep finding ways to provide it with more commodities at lower costs—and those *ways* sought by the economy and its society amount to *technique*: the most efficient way to produce something for the lowest cost at a given point in time. The long-term effect is that the economy and its society become addicted to *technique*'s efficiency because *technique* is the sole means by which the cycle of inexpensive production and consumption at every level of society not only keeps going but also keeps growing. Resonances with Marx are strong here and I will show how Ellul's theorization of *technique* aligns with and differs from Marx's theorization of *capital*. For now, though, I will continue with a summary of the basic character and effects of *technique* before moving to Ellul's reworking of *capital*.

Returning to the mutation of *tékhnē*, *poiēsis*, and *epistēmē*, the sequencing of the shift is as follows: with the Industrial Revolution comes the modern world's idealization of efficiency, and with the modern world's idealizations of efficiency, comes *technique*: thinking (*epistēmē*) about the most efficient way (*technique*) to produce the highest possible yield of something for the lowest possible expenditure of money, time, energy, and so on when acting (*poiēsis*) or making (*tékhnē*).[6] Then, knowing, acting, and making in the modern world become conflated with *technique* to the point of becoming eclipsed by it. This is because the demands of efficiency increasingly pervade every aspect of society, including a person's day-to-day life. Ellul initially specifies in *Technological Society* that *technique* has five characteristics—automatism, self-augmentation, monism, universalism, and autonomy—but given he adds to the characteristic s in his later texts, and given these characteristics have been adequately summarized elsewhere, the determinative value of efficiency is *technique's* most important quality to consider.[7] Again, the example of the painter's *technique* is demonstrative.

In the seventeenth century, Rembrandt made his own paint before composing his works. This is classic *tékhnē* because it represents Rembrandt's unique *technique*. In the twentieth century, Jackson Pollock purchased cans of ready-made house paint and splashed their contents on a canvas. This is modern *technique* because it is relatively reproducible when compared to Rembrandt's method. Now, in the twenty-first century, many artists forgo physical paint and canvas altogether and use digital media that bypasses the work involved in using paints, brushes, and canvases.[8] *Technique* consequently shifts from Rembrandt's elegant brushstrokes and handmade paints, to Pollock's furious-flicking of store bought paints, to the digital designs of graphic artists. That is to say, the most efficient way to produce the highest yield for the lowest possible expenditure increasingly determines the artist's process of making.[9]

Or, modern *technique* significant alters classic art (*tékhnē*). Note, this is not as much a judgment of aesthetic value—although it is that too given Ellul's severe criticisms of modern art—so much as it is a description that concerns ethics, or, how classic *tékhnē* mutates into modern *technique*.[10] Ellul argues that this mutation signals a capitulation to an increasingly technologized world that values efficiency more than any other values.[11] Moreover, to clarify, Ellul's point is not that efficiency itself is categorically problematic. His point is that if *technique* makes efficiency an absolute value and excludes other possible values, then *technique* will drastically, and in many cases dangerously, alter how ethics are perceived and articulated in the modern world.[12] In Marxist terms, his concern is the "overdetermination" of *technical* efficiency in the modern world.[13] For this reason, his use of the word *technique* instead of *tékhnē* linguistically signals his contention that, when we speak about ethics today, we can no longer speak about them using classic terms such as *tékhnē*, *poiēsis*, and *epistēmē*. Rather, when we speak about ethics now, we should speak about modern *technique* and its overdetermined value of efficiency, because it pervades thought, action, and making at every level of society.[14]

Although tacit understandings of and experiences with technologies make it tempting to focus on the various manifestations of *technique*—machines, vehicles, devices, methods, and so on—and conclude that what *technique* produces—that is, what the rational process of *technique* yields in its material form—is synonymous with *technique*. But this detrimentally privileges a study of the physical manifestations of *technique* over a study of the rational process of *technique*.[15] Of course, some manifestations of *technique* are visible and tactile (cars, phones, computers, furniture, cooking utensils, etc.). Yet others are invisible and untouchable (speech performance, organizational strategies, self-help programs, meditation or prayer practices, and so on). In fact, Ellul aims to expose the deeper thoughts and actions that precede the production of this or that instance of *technique* because it is the deeper thoughts and actions, which hold the most power in the rational process of *technique*, and consequently, hold the most power at every level of society. Examining Ellul's reworking of Marx's theorization of *capital* illustrates how this is possible.

Marx argued that the way *capital* changes the production process of a worker is enough to make a table metaphorically stand on its head and literally shift societal structures in novel and problematic ways. Similar to Ellul who writes about *technique's* power in the twentieth century, Marx hints at mutations in classic *tékhnē* when he describes how *capital* powerfully alters production and consumption processes in the nineteenth century:

> It is as clear as noon-day, that man, by his industry, changes the forms of the materials furnished by Nature, in such a way as to make them useful to him. The

> form of wood, for instance, is altered, by making a table out of it. Yet, for all that, the table continues to be that common, every-day thing, wood. But, so soon as it steps forth as a commodity, it is changed into something transcendent. It not only stands with its feet on the ground, but, in relation to all other commodities, it stands on its head, and evolves out of its wooden brain grotesque ideas.[16]

The table's commodification consequently alienates the carpenter from his or her work, and the more this process occurs, the more socio-economic relationships at individual and structural levels are altered. So, Marx continues, describing how the commodification process affects the laborer: "In [the commodity] the social character of men's labor appears to them as an objective character stamped upon the product of that labor."[17] Which is to say, what the laborer put into the production of the commodity appears to be an observable part of it. But it only appears so, because, "It is different with commodities. [In fact], the existence of [things as] commodities, and the value relation between the products of labor which stamps them as commodities, have absolutely no connection with their physical properties and with the material relations arising therefrom."[18] Essentially, the table no longer is the product of one laborer's work, but multiple laborers; the tools and materials used to make the table no longer are owned by the laborer(s), but the capitalist; the table no longer is made to be used, but sold; and finally, the system of circulating *capital* that is required to produce, distribute, sell, and purchase the table is built, run, and bolstered by the capitalist, all to the exclusion of the laborer. The net result is prosperity for the capitalist and the bourgeois, and alienation for the laborer and the working class. And for Marx, this entire production and consumption process is what makes a commodified table stand on its head and alter socio-economic relations in the nineteenth century.[19]

The critical move Ellul makes in relation to Marx's theorization of *capital* as summarized here is an expansion of its effects to *all* areas of life (social, economic, political, cultural, religious, etc.) because *technique* makes it possible for *capital* to expand to all of these areas. According to Ellul, nineteenth-century *capital* might have produced *technique*, but in the twentieth century, *capital* needed *technique* for growth into new markets. Consequently, *technique* (the most efficient way to produce something at a given point in time) drives *capital* (what is produced at a given point in time), making *technique* the only possible means for *capital* accumulation and circulation.[20] Of course, *technique* and *capital* still need each other. *Technique* needs the investment of *capital* to be actualized in the world, but *technique* always holds the power because it is the only way *capital* can expand to control new territory.[21] Note, Marx's focus on ownership of the *means* of production is especially critical here, because Ellul is able to show how *technique*, as the sole *means* for efficient production, starts to control both (1) the people who are owned by the capitalist and (2) the capitalist who becomes increasingly determined—owned, in fact—by his or her own means of production. Altogether, it is the power of *technique's* overdetermined criterion of efficiency to alter not only the production of the working class carpenter, but also the administrative practices of the middle-class-factory floor manager, and the organizational strategies of the factory's upper-class chief executive. The net result is universal—societal, and then global—alienation. Hence, Ellul's thesis for *Technological Society*: How would Marx designate *capital* differently were he writing in 1940?[22] A substantial part of Ellul's answer is *la technique*, but that is only part of the answer. The other part is *la technologie*, the discourse that justifies *la technique*. Together, *la technique* and *la technologie* produce *la propagande.*

Technology as a Constant Discourse

La technologie differs from, but is tied to *la technique*, because "technology is discourse on technique."[23] It involves how people speak and write about *technique*, how people study *technique*, and how people learn to apply *technique*. Or, in different terms, *technology* is the way people are indoctrinated into the rational thought processes of *technique*.[24]

On the one hand, Ellul does not offer an entirely new definition of *technology*. It is like the words *theology*, *philosophy*, *anthropology*, *ecology*, *geology*, and so on. Each of these words is used to describe a discourse on *theos*, *philos*, *anthropos*, *ecos*, *geos*, and the like.[25] In the same way *geology* describes *geos*, *technology* describes *technique*. On the other hand, Ellul identifies something unique about *technology*. It is a kind of discourse that unequivocally and univocally justifies *technique*. So, whereas a skilled geologist learns from *geology* to study a rock formation from various vantage points, an effective technologist learns from *technology* to study and apply *technique* from a single vantage point: the one efficient way to yield the most with one's action and making.[26] For this reason, *technology* is an efficient, but constricting discourse that can powerfully determine a person's thinking, acting, and making.

Ultimately, Ellul's concerns about *technology* concern language. If *technique* is a rational thought process, and people think in language, then *technology* is the way people have their thinking, acting, and making significantly determined by *technique*. Such is the power of *technology*. If it did not exist, neither would *technique*, because the only way a person can learn *technique* is through the discourse—the language—of *technology*. In fact, Ellul thinks *technology* is powerful enough in the modern world to produce an ideology of *technique* that inspires religious devotion. As he writes in *Technological Society*, "Technique, then, brings its own ideology; and every technical realization engenders its own ideological justifications."[27] However, "Not technique but man is responsible." Since, "[Man's] works involve immense sacrifices, it is necessary to justify the sacrifices (a fact I shall return to in my study of propaganda)."[28] Consequently, "Man creates for himself a new religion of a rational and technical order to justify his work and be justified in it."[29] I have left Ellul's comment about returning to his study of *propaganda* in this quote to indicate the relationship between *technique*, *technology*, and *propaganda*, the latter of which I will feature in the next section. Note here, however, the religious language appearing as it did in the passages referenced in the previous chapter. The added nuance in this quote is that Ellul says more about how the idol of technology is constructed in the modern world. That is to say, *technology* does not just teach *technique* in an applied sense, but it also justifies *technique* in a religious sense, and this justification creates a religion of a rational and technical order.

With the above in view, the simplest way to summarize the relationship between *technology* and *technique* is as follows: as *technology* produces an ideology of *technique*, it compels people to idolize *technique*. The passage just quoted is from the early sociological text *Technological Society*, but Ellul makes similar points in early theological texts such as *Presence in the Modern World*.[30] He will argue these points until the end of his life.[31] For this reason, whether Ellul writes from a sociological-historical or theological-ethical perspective, the way he describes *technology*'s justification of *technique* is similar to the way Isaiah 44:13–17 breaks down the construction of an idol. For there we read the following:

> The carpenter stretches a line, marks it out with a stylus, fashions it with planes, and marks it with a compass; he makes it in human form, with human beauty, to be set up in a shrine. He cuts down cedars or chooses a holm tree or an oak and lets it grow strong among the trees of the forest. He plants a cedar and the rain nourishes it. Then it can be used as fuel. Part of it he takes and warms himself; he kindles a fire and bakes bread. Then he makes a god and worships it, makes it a carved image and bows down before it. Half of it he burns in the fire; over this half he roasts meat, eats it and is satisfied. He also warms himself and says, "Ah, I am warm, I can feel the fire!" The rest of it he makes into a god, his idol, bows down to it and worships it; he prays to it and says, "Save me, for you are my god!"[32]

What makes the idol an idol in this passage? Is it the carpenter's work (*technique*) or the carpenter's words (*technology*)? Both, of course. But what if the question was framed differently, and it is asked: What gives the idol its power? Then the balance of power shifts to the carpenter's words; the words that determined his thoughts about the wood's use; and, the words he spoke over the wood when the idol was fashioned and the fire burned.

Read through a biblical frame, the technical phenomenon Ellul describes in these passages is similar to the phenomenon described in the passage from Isaiah. That is, *technology* is the discourse that can turn *technique* into an idol. The difference between the two cases is that Ellul describes the phenomenon in its modern setting, a context with idols that differ in kind and scale from the carpenter's in Isaiah 44. Here is an example from *Technological Society*: "The technician uses technique perhaps because it is his profession, but he does so with adoration because for him technique is the locus of the sacred. There is neither reason nor explanation in his attitude."[33] Yet, "The power of technique, mysterious though scientific, which covers the whole earth with its network of waves, wires, and paper, is to the technician an abstract idol which gives him a reason for living and even for joy."[34] Like the carpenter, the technician worships the products of his or her *technique*, but what the carpenter worships and the technician worships differ in scale. The waves, wires, and paper sacralized by the technician cover much more territory than that of the wood sacralized by the carpenter. So, it is the idol's scale, as much as the idol itself, that worries Ellul.

It was Ellul's experiences leading up to and during the wear, which gave ground to his worries about *technology*'s idolization of *technique*. In a late interview, he describes attending a Nazi rally in 1935, in Munich, after being invited by Christian associations to attend. He recalls how effectively a mass gathering of people could be manipulated and controlled by the basic technologies used there. When asked if the experience influenced his work, Ellul replies: "Absolutely. It was fascinating to see how easily a crowd could be whipped up and, welded into a single unit . . . . No-one, absolutely no-one, had any individual reactions left."[35] After experiences like these, Ellul will warn in texts like *Humiliation of the Word*, "When [language] uses a loudspeaker and crushes others with its powerful equipment, when the television set speaks, the word is no longer involved, since no dialogue is possible. What we have in these cases is machines that use language as a way of asserting themselves. Their power is magnified, but language is reduced to a series of sounds which inspire only reflexes and animal instincts."[36] For this reason, Ellul's concerns are not limited to the ways *technology* idolizes *technique*. Although such concerns are valid in and of themselves, it is important to emphasize that Ellul's concerns extend to the effects of *technology's* idolization of *technique*. For, when the idol is constructed, it can provide a kind and scale of power that Ellul describes in these passages: power that enchants people, crushes people, and transforms people into something more theriomorphic than human. And it can do all of this on a mass scale.

Although these observations are more sociological than theological, Ellul is making three illuminating points that have implications for his sociology, theology, and ethics of technology. The first illuminating point has to do with why people find idols worthy of worship: they promise power, and if they did not, it is unlikely they would be considered worth worshipping. The second illuminating point has to do with how idols are constructed: they are made in the mind and with the hands, but they are idolized by language. What we say (and write) can sacralize (or desacralize) our thoughts and the labor of our hands. The third illuminating point has to do with idolatry and self-deception: it is easy to assume that we are exempt from the kind of idolatry described here, but Ellul does not allow us to make such assumptions easily. He frequently points out that the abundance of *technological* discourse in the modern world implies that many of us idolize *technique*, and consequently are seduced, crushed, and transformed by *technique*, without us ever being fully conscious of it. Altogether, Ellul's point about *technology* is we ought to speak about *technique* carefully, as it may be that we have been seduced to justify, sacralize, and idolize it because power is attractive, and for that reason, we do not wish to think or speak about it in a critical way.

But as powerful as *technology* and *technique* are, they do not contain enough power on their own to whip up crowds and crush people in the ways Ellul describes in these passages. In order for the idol to scale its power, something else is needed, and Ellul says this something else appears when *technology* and *technique* produce *propaganda*. For this reason, the next section focuses on Ellul's understanding of *propaganda* and its relationship to *technique* and *technology*.

Propaganda, Produced by Technique and Technology

As *la technique* and *la technologie* power each other in the modern world, *la propagande* is produced, and it scales the power of both. Metaphorically, if *technique* is dust and *technology* is wind, then *propaganda* is the dust storm tearing through modernity. Literally, *propaganda* is the result of *technique* and *technology*'s dialectical interaction and reinforcement in the modern world. Ellul writes, "Not only is propaganda itself a technique, it is an indispensable condition for the development of technical progress and the establishment of a technological civilization. And, as with all techniques, propaganda is subject to the law of efficiency."[37] Simply put, *propaganda* is the *technique* that can make an entire civilization *technological*. To show how this is possible, Ellul describes two form of modern *propaganda*: political and sociological.

Political *propaganda* is what most people think about when they hear the word *propaganda* used. It "involves a technique employed by a government, a party, an administration, a pressure group, with a view to changing the behavior of the public."[38] It acts vertically on society as it is deployed from technicians to societies. It aims for immediate social engineering.[39] The second type of modern *propaganda* is sociological. It is less conspicuous than the political kind because it acts in a horizontal, holistic way in societies. It amounts to "the group manifestations by which any society seeks to integrate the maximum number of individuals into itself, to unify its members' behaviors according to a pattern, to spread its style of life abroad, and to impose itself on groups."[40] Or, whereas political *propaganda* aims to produce immediate actions, sociological *propaganda* aims to produce conformity to certain thoughts, beliefs, behaviors, and ways of life. Even simpler, sociological *propaganda* molds and political *propaganda* jolts societies.

Just as the prevalence and power of *technique* and *technology* in the modern world make it difficult to discern its ideological and in some cases idolatrous hold on a society, both forms of *propaganda* are likewise so common that they cease to be identified as such. Ellul writes, "The combination of advertising, public relations, social welfare, and so on produces a certain general conception of society, a particular way of life. From then on, the individual in the clutches of such sociological propaganda believes that those who live this way are on the side of the angels, and those who don't are bad." Ultimately, "It is a matter of propagating behavior and myths both good and bad. [Such] propaganda becomes increasingly effective when those subjected to it accept its doctrines on what is *good* or *bad* (for example, the American Way of Life). There, a whole society actually expresses itself through this propaganda by advertising its kind of life."[41] As in many other of Ellul's sociological analyses of *propaganda*, religious language appears. Here, Ellul says *propaganda* uses "myth" to form doctrines of good and bad, and these myths and their doctrinal formulations really influence a society's perspectives, beliefs, and actions. The importance of *myth* is elaborated in the next chapter, but for now, the fundamentally, and dangerously formative character of political and sociological *propaganda* is important to examine in this section.

Before proceeding any further with Ellul's analysis of *propaganda*, however, it is essential to offer a word of caution. As Ellul observes, "It is [very] difficult to determine what constitutes propaganda in our world and what the nature of propaganda is. This is because it is a *secret* action."[42] In other words, however conspicuous or inconspicuous *propaganda* appears to us, it is dangerous for us to assume that either (1) "everything is propaganda," and therefore dismissively conclude that "nothing" is *propaganda*, or (2) "abandon the term propaganda altogether" and study the phenomenon as if it is not as threatening as it really is. The danger with both responses is that we would be persuaded to conclude that we are free to make, disseminate, and be molded by *propaganda* without it having real, adverse effects on ourselves and other people.[43] Actually, Ellul contends, as do I after reading Ellul's work on *propaganda*, not everything is *propaganda*, we should not abandon the term, and we should expose and resist it even if it has become so commonplace and effective in the modern world that most people are oblivious to its existence and power. In fact, Ellul argues, it is the intellectuals of society who are most easily persuaded by *propaganda* because they trade in ideas, and modern *propaganda* tends to use the most appealing and seemingly avant-garde ideas in a society to manufacture powerful ideologies.[44]

Context is crucial to explain why Ellul believes his equation of sociological *propaganda* with advertising, public relations, and social welfare is warranted. What arrests his attention with postwar phenomena such as the kind he describes above is the undeniable uniformity, at the level of discourse, between the agitative political *propaganda* everyone envisions when the word *propaganda* is used, and the commonplace sociological *propaganda* present in advertising, public relations, and social welfare. Even earlier than *Propaganda*, Ellul convincingly argues this point in *Technological Society* when he observes, "Compare *Life* magazine with the *Soviet News* and you will see that 'the good of humanity' is conceived in almost identical terms in the United States and in the Soviet Union; the difference lies mainly in the persons charged with securing it. In both cases, the social good can be reduced to a few concrete and precise factors."[45] Ellul was infuriated by this uniformity because the *propaganda techniques* and *technologies* used to form Nazis and fascists were the same *propaganda techniques* and *technologies* being used to form consumers with notions of the good life. Furthermore, Ellul was concerned because he knew that this kind of *technical* and *technological* engineering with *propaganda* could quickly shift from being technologies used to make consumers and sell commodities, to technologies again being used to make soldiers and start wars. And unarguably, history has proven these concerns to be well-founded. Using *propaganda techniques* to produce loyal consumers and well-behaved citizens can seem harmless to those who are doing the producing, but once the *propaganda* machine starts, it becomes difficult to stop it from being accelerated by a faction (or factions) in a society and used for more extreme, dangerous purposes. Again, history and the present has shown that such acceleration happens frequently.

One postwar example among many verifies Ellul's analysis of modern *propaganda* and his concern about it. It also demonstrates that Ellul was not the only person in postwar France who noticed a similarity between modern advertising and *propaganda*. Tony Judt describes France's initial reaction to the appearance of Coca-Cola in Europe as follows: "Between 1947 and 1949 the Coca-Cola company opened bottling plants in the Netherlands, Belgium, Luxembourg, Switzerland, and Italy. Within five years of its creation West Germany would have 96 such plants and became the largest market outside the US itself."[46] Although, "Some voices had been raised in protest in Belgium and Italy, it was in France that Coca-Cola's plans unleashed a public storm."[47] So much that, "When *Le Monde* revealed that the company had set a target of 240 million bottles to be sold in France in 1950, there were loud objections—encouraged but not orchestrated by the communists, who confined themselves to the warning that Coke's distribution services would do double duty as a US espionage network."[48] As a result, "Le Monde editorialised on 19 March 1950, 'Coca-Cola is the Danzig of European culture.' The furore over 'Coca-Colonisation' had its light side" of course "but the sentiments underlying it were serious."[49] Here, Ellul's analyses of the uniformity and ambivalence *technique*, *technology*, and *propaganda* are given substance. Coca-Cola, espionage, statecraft, and *propaganda* were uniformly and ambivalently connected, and reliant upon each other, for development and operation.

From a postwar French perspective, things like Coca-Cola represented more than an advertised soft drink. They signified a *propaganda* enterprise, the imperialist ooze of the American Dream carbonated in compact containers. Yet, from a postwar American perspective, Coca-Cola represented an aspect of everyday life as common as the cheeseburgers and automobiles that appeared in their advertisements. So which is the correct interpretation—the French or the American? Ellul helps readers to see that the correct answer to the question is: *both!* A deeper analysis of the discourse demonstrates that advertisements, Coca-Cola, hamburgers, and automobiles all represent the American way of life as much as *propaganda*, espionage, imperialism, and dropping bombs. Furthermore, as time passes, Coca-Cola and the way of life it installs through *propaganda* gradually integrates itself into the French way of life. After a period of initial fury subsides, life simply carries on in France as if Coca-Cola and what it first represented never was offensive or threatening. At that point, the *propaganda-advertising* campaign has achieved its mission. The networks needed for statecraft and espionage are set up, and put to use, and everyone is happy drinking their made-in-America soft drinks.

Ellul also clarifies that *propaganda* is not unidimensional, as if propagandists are solely responsible for the social phenomenon. Rather, Ellul contends that propagandees also ought to be held responsible because they tend to welcome, even desire *propaganda*'s appeal to and meeting of deep human needs. Which is to say, *propaganda* is defanged when it ceases to appeal to its audience, so it cannot create ideas, beliefs, and stories out of nothing. Instead, to be effective, *propaganda* has to engage and inflate what is already present in its audience. Two of the primary ways this engagement and inflation happens are *myth* and *justification*. First, speaking about the relationship between *propaganda* and *myth*, Ellul writes, as a "*remedy for loneliness* [propaganda] corresponds to deep and constant needs, more developed today, perhaps, than ever before: the need to believe and obey, to create and hear fables, to communicate in the language of myths."[50] Second, Ellul writes about *propaganda* and its furnishing of justifications for people, "To the extent that man needs justifications, propaganda provides them. But whereas his ordinary justifications are fragile and may always be open to doubts, those furnished by propaganda are irrefutable and solid. The individual believes them and considers them to be eternal truths. He can throw off all sense of guilt."[51] Again, such passages prefigure later chapters, because they indicate how Ellul can move a sociological analysis of *technique*, *technology*, and *propaganda* into existential territory, and, in turn, move this existential analysis into theological and ethical territory. The existential, which is always intertwined with the social, is a meeting point for Ellul's dialectic of sociology and theology, and it is belief, obedience, *myth*, justifications, eternal truths, and senses of guilt, which stand out in these passages as the existential zones that can be written about from two perspectives—sociological-historical and theological-ethical—rather than one. But these aspects of the passages just quoted can be put aside for now. Here, the primary concern is Ellul's understanding of *propaganda*, how it can make an entire civilization *technological*, and what all of this has to do with idolatry.

If the scripture passage from Isaiah 44 referenced in the previous section is recalled here, what Ellul is up to when he describes certain aspects of the *technological* phenomenon in some of his most well-known texts becomes quite clear. For example, in *Technological Society*, he describes the sacralization of modern technologies in the following way, and he is worth quoting at length given the passage's thematic significance:

> One sign, among many, of the feeling of the sacred that man experiences in the face of technique is the care he takes to treat it with familiarity. Laughter and humor are common human reactions in the presence of the sacred. This is true for primitive peoples; and for the same reason the first atomic bomb was called "Gilda," the giant cyclotron of Los Alamos "Clementine," the atomic piles "water pots," radioactive contamination "scalding." The technicians of Los Alamos have banned the word atom from their vocabulary. These things are significant. In view of the very different forms of technique, there is no question of a technical religion. But there is associated with it the feeling of the sacred, which expresses itself in different ways. The way differs from man to man, but for all men the feeling of the sacred is expressed in this marvelous instrument of the power instinct which is always joined to mystery and magic. The worker brags about his job because it offers him joyous confirmation of his superiority. The young snob speeds along at 100 m.p.h. in his Porsche. The technician contemplates with satisfaction the gradients of his charts, no matter what their reference is: it is the common expression of human power without which they

> would find themselves poor, alone, naked, and stripped of all pretensions. They would no longer be the heroes, geniuses, or archangels which a motor permits them to be at little expense.[52]

Whether it is an atomic bomb, a cyclotron, or a fast car, *technology* is the means by which a person's experiences and perceptions of the sacred are transferred onto *technique(s)*. *Propaganda* is the means by which the sacralization scales to the size of an entire society. The carpenter in Isaiah 44 and the scientist and young snob in this passage sacralize objects produced by *technique* in the same way, but the scale of the sacralization and its threats differ significantly. In this passage, Ellul is showing how the phenomena of transferring sacred significance to bombs, cyclotrons, and cars are not isolated, but related, just as Coca-Cola, espionage, and statecraft were related in the example provided earlier. Although the *technical* objects differ in kind and power, the action of ascribing sacred significance to the objects is identical. Language, the way in which the discourse of *technology* hardens into *propaganda* and exerts pressure on a society, is what makes the process identical. Essentially, the sum of the technological idolatry is represented by the parts described in this passage. The scientist and the young snob are parts of the whole societal sacralization of *technique* through *technology* and *propaganda*, and this societal sacralization of *technique* is dangerous when people start to worship and laugh as they stand in front of an atomic bomb. In fact, as a historic case in point, Robert Oppenheimer, leader of the Manhattan Project, famously approached the first atomic explosion as a quasi-divine event. He quoted Vishnu in Bhagavad Gita before detonating the bomb: "Now I am become Death, the destroyer of worlds."[53] Moreover, the first atomic explosion was named the *Trinity Test*. Were these passages quoted and atom bomb tests named with religious earnestness or sickening senses of humor? If Ellul's analysis is correct, they were quoted and named with a mixture of both, because the idolatry of technology, like all kinds of idolatry, combines religious earnestness with sickening senses of humor. And this, as Ellul shows in his work, is incredibly dangerous.

With the above in view, the *technical* objects in a society may vary, but the action of sacralization learned and habituated in society through *technology* and *propaganda* is universal and uniform. *Technique* and *technology* ensure uniformity. *Propaganda* ensures universality. Most importantly for later chapters in this book, Ellul often says that this uniformity and universality is achieved through manipulations of language, and, in particular, the dissemination of discourse that provides people with justifications for their thoughts, beliefs, words, and actions—or, more simply, provides people with ways to self-justify their existence, not just in a spiritual sense, but in a spatial and temporal sense also. For example, Ellul writes the following in *Propaganda*:

> An analysis of propaganda therefore shows that it succeeds primarily because it corresponds exactly to a need of the masses. Let us remember just two aspects of this: the need for explanations and the need for values, which both spring largely, though not entirely, from the promulgation of news. Effective propaganda needs to give man an all-embracing view of the world, a view rather than a doctrine. Such a view will first of all encompass a general panorama of history, economics, and politics. This panorama itself is the foundation of the power of propaganda because it provides the justification for the actions of those who make propaganda; the point is to show that one travels in the direction of history and progress. That panorama allows the individual to give the proper classification to all the new items he receives; to exercise a critical judgment, to sharply accentuate certain facts and suppress others, depending on how well they fit into the framework.[54]

In literal and metaphorical ways, *propaganda* and *technology* are the means by which the idol of *technique* is constructed and justified from the individual to the societal level. Taking the example of *The Trinity Test*, one might ask: How do the people performing such tests and the people viewing the tests live or on the evening news justify to themselves such dangerous, potentially world-ending explosions of *technical* power? Ellul's answer is: *technology* and *propaganda*, the way an entire society sacralizes *technique* and discursively justifies that idolatry with collectively agreed upon ideals of history and progress. In later chapters, the theological and ethical facets of this discursive sacralization and justification will be brought into view. At this point, however, it is important to turn to Virilio's philosophy-phenomenology of modernity's technological idolatry to identify where it aligns with and differs from—and consequently reinforces—Ellul's work as it has been described so far.

War and Idolatry

Virilio's analyses of the construction, sacralization, and justification of technological idols are not as nuanced as Ellul's sociological-historical analyses of *technique*, *technology*, and *propaganda*. However, his vivid philosophical-phenomenological analyses of the effects technological idols can have on the environment, including the bodies that exist in and which make up the environment, often exceed Ellul's nuance in this realm. To reiterate here the primary difference between Ellul and Virilio's methods for identifying, exposing, and dismantling technological idolatry in the modern world, Ellul focuses on language and justification, which is rooted in his Protestant Reformed tradition, and Virilio focuses on mediation and the senses, which is rooted in his Catholic tradition. Keeping this in mind, it can be said that

Ellul reforms *tékhnē*, *poiēsis*, and *epistēmē* with his sociological-historical analyses of *technique*, *technology*, and *propaganda*, and Virilio reframes these terms with philosophy-phenomenology that pulls the reader's focus toward the environment (*phusis*) in which all human thought, action, and making takes place.

As Virilio turns the reader's attention toward the environment and bodies affected by technological idols, he shows how technological idols increase speed (*drómos*), harness motion (*kīnēsis*), and manipulate light (*phōs*) in powerful, often damaging ways. Here, the well-known facets of Virilio's work come to the fore, particularly his studies of war and speed. As Virilio says an interview, "War has always been the laboratory of the future."[55] It is "the laboratory of speed" and "the laboratory of modernity, of all modernities."[56] Consequently, he says:

> War is not merely an amoral phenomenon, it is an experimental phenomenon inasmuch as it reverses productivity relations. War produces accidents. It produces an unheard-of accident, which is upsetting the traditional idea of war. Substance is necessary and accident is contingent and relative! That is the traditional story of the return to the accident. In war time the opposite is true. Here accident is necessary and substance relative and contingent. What are war machines? They are machines in reverse—they produce accidents, disappearances, deaths, breakdowns. I think war in this sense conveys something which at present we are experiencing in peacetime; the accident has now become something ordinary.[57]

Resonances of Aristotle and Nicholas of Cusa are strong here, just as they are in other places Virilio's work. Such resonances receive attention in a later chapter, but for the moment, the way these motifs connect to technological idolatry and Ellul's work require elaboration. To that end, Michael Hirsch helpfully describes the intricate connections between Virilio's focus on war, machines, the concept of deus ex machina, speed, accidents, and idolatry—most of which appear in the passage just quoted. As Hirsch writes, "When machines begin to be idolized, apotheosis has occurred. Accordingly, the concept *deus ex machina* becomes simply *machina deus*: the machine is no longer a vehicle through which a god appears or is revealed, but rather the machine *is* god, the machine incarnates god, hypostasizes God."[58] On top of a good characterization of how Virilio exposes technological idolatry with studies of war, Hirsch's analysis also captures how Virilio skillfully exposes the strange, often dangerous reversals caused by technologies. In fact, in the Virilio passage quoted earlier, one can see such reversals at play when Virilio describes war's reversals of accident and substance relations. This is why Hirsch is correct to suggest that Virilio's philosophy-phenomenology of technology often reveals how "this new god, the machine-god, displays certain characteristics, which at times betray the accident of [a] divine transformation."[59] In other words, Virilio's preoccupation with the accident(s) caused by the substance(s) of technologies is one of his primary means for exposing, and attempting to dismantle, the technological idols constructed, sacralized, and justified in the modern world. As he focuses the reader's attention on technological accidents pioneered and used in and beyond the theatre of war—in the broadest and deepest senses the word *accident* can be taken—Virilio constantly shows readers that there is something in the substance of what we think, do, and make that inclines us to worship power, make idols, and enact violence. The theological term Virilio uses for "this something" in the substance of what we think, do, and make, is "sin," but he does not use it often in his work.[60] Instead, "accidents" for Virilio is an idiom that allows him to cover a great deal of analytical territory, including that of sin. More will be said on this in the next chapter.

With the aforementioned in view, "war," as Virilio frequently illustrates, is basically the word most people use to describe the times and places that so many of the powerful idols of the future are designed, built, put to use, worshipped, and propagated. For example, Virilio writes, "The ecstasy of the crewman of the *Wisconsin* and *Missouri* at the takeoff of the Tomahawk [missiles], those oh-so-'smart' weapons equipped with inertial guidance systems and terrain-following radar that directed them infallibly to the end of their course, the final explosion more than a thousand kilometers away," transforms the crewmen into "unrepentant fetishists marveling at the remote manipulation of some wax or wooden figurine, a magic rite whose object is to direct an occult force toward a chosen act or the distant destruction of beings and things, thanks to the diversion of their digitalized image or of their radar ghost."[61] Virilio concludes that this scene literally and metaphorically represents "the reconstitution of an old power ritual that comes from who knows where—maybe the age-old origins of an *applied science* that believes it is beyond suspicion and merely returns to it."[62] Note especially Virilio's mention of "applied science" here, as it reframes notions of classic *scientia*, the Latin word for *knowledge*, which has a place in the family of Greek *epistēmē*, the word for knowledge of theoretical truth and the etymological root of *science*, *thought*, and *knowing*. As it has been said throughout the ages, *scientia potentia est (knowledge is power)*, and Virilio constantly reminds readers that the *scientia* of war is only an extreme, violent instantiation of the everyday, uniform, and universal kinds of human thought that is prone to worshipping power, making idols, and damaging ourselves, other people, and the world we are in. As Virilio observes, "All of us are already civilian soldiers, without knowing it. And some of us know it. The great stroke of luck for the military class' terrorism is that no on recognizes it. People don't recognize the militarized part of their identity, of their consciousness."[63] It is this uniformly and universally "militarized," often idolatrous way of thinking (*epistēmē*), acting (*poiēsis*), and making (*tékhnē*), that Virilio works to identify, expose, and resist with his philosophy-phenomenology. By exposing the militarization of our consciousness and identity, resisting the idols that this consciousness and identity tends to construct and worship, Virilio aims to demilitarize our militarized societies and dismantle the sacralized technologies that these kinds of societies tend to make and worship.

If Virilio's account of technological idolatry seems objectionable, a recent example from 2017 confirms his phenomenological descriptions of Tomahawk missile worshippers aboard the *Wisconsin* and *Missouri* ships. North America's liberal media, which claimed it was "the resistance" to former president of the United States Donald Trump, also unreservedly stated, he became "presidential" the moment he ordered a Tomahawk missile strike on a Syrian military airfield. As footage of the missiles being fired in the night from the decks of US Navy Destroyers was released by the Pentagon and repeatedly featured with favorable commentary on so-called resistance news stations like MSNBC, anchor Brian William expressed shock, awe, and praise as he watched clips of the smart rockets blazing in the sky: "We see these beautiful pictures at night from the decks of these two U.S. Navy vessels in the eastern Mediterranean" and "I am tempted to quote the great Leonard Cohen: 'I am guided by the beauty of our weapons.' "[64] Indeed, *scientia potentia est*, and we are prone to worshipping power and idolizing the technologies, which seem to promise it. As we do that, the phenomenon we call *war* is simply the time and space when and where the latest technologies of power are invented, tested, and eventually given to the masses for adoration and application. The internet, as Virilio often notes, is an example of this phenomenon, which, as it happens, drove the advance, and transformation, of history.[65] The internet was invented to keep communication lines running in the event of atomic war, and in this sense, Virilio points out, it was considered a means of salvation by the people who invented it and first put it to use. Then, as time passed, the Cold War drew to a close, the technology was made available to everyday citizens, and it was—just as it still is—ascribed sacred value by people who worship the power an everyday technology like the internet appears promise. And herein lies the danger, Virilio warns: As much as we try to suppress the substance of militarization (i.e., violence) contained within common technologies like the internet, accidents—not mistakes, but unique qualities of a particular substance—always appear to reveal to us that we cannot, at least not with our own intelligence, will, and strength, suppress the militarized substance contained in what we think about, do, and make.[66] It is for this reason that Hirsch notes, "To the latin ear[,] *machina*" an essential part of the deus ex machina, "Would generally suggest a siege-engine, hence, a war machine. [And] this connection becomes all the more important when we consider Virilio's abundant work on the military-industrial complex."[67] That is to say, a broad and deep look at Virilio's oeuvre with these connections between war, technology, and idolatry in mind shows how a significant portion of his work in these areas can be read as an attempt to expose, dismantle, and reverse the negative effects of modernity's idolatry of technology. Or, to put it differently, when Virilio writes about war, when he dwells in this abominable, accident ridden laboratory of war he is often declaring and waging war on technological idolatry.

Hirsch has done an excellent job summarizing important connections in Virilio's work on technology, so I will not repeat his work. Instead, I will build from his helpful observations, and begin by offering the simple remark that I think it is strange that few scholars engage with these important aspects of Virilio's work, especially when one considers how direct Virilio can be about technological idolatry in both interviews and writing. For example, in *Negative Horizon*, he writes:

> With the advent of nuclear arms, the old theological doctrine of the *just war* that associated state powers and the Vatican for so long has become obsolete. The proof of this is the journey to Hiroshima after Auschwitz of the one who declared in 1980: "Nuclear holocaust threatens. In whose political interest could there ever be a new war?" If Pius XI denounced in his time the statolatry of a rising Fascist state, Jean Paul II persisted in denouncing terrorism to the United Nations, in Ireland, and elsewhere, the cult of Peace by deterrence, the idolatry of an absolute weapon, supposedly capable of bringing about salvation. It is thus evident that if war tends to be definitely rejected by the Church, it is because it has attained such proportions as to have gained the power to obscure divine revelation: *the scientific and technological apocalypse now [screens] the mystical apocalypse.*[68]

Reasons for this absence of engagement notwithstanding, this passage offers a glimpse into the numerous ways Virilio exposes the effects technological idolatry and other forms of idolatry can have on the environment, bodies, and senses in which we do our thinking, acting, and making. Specifically, in this passage, he observes that the idolatry of atomic technology contains the power to obscure divine revelation and cause a mystical apocalypse that does not reveal a kind of substance we should wish to be revealed. Instead, the atomic apocalypse promises to disappear everything and everyone in a nearly instantaneous explosion of technological power. These are the kinds of ambitious, perhaps prophetic observations frequently developed in Virilio's work. In this case, such observations show that the technologies we idolize—because we think and believe they will save us—tend to be technologies that contain enough power to destroy us, other people, and our environment. And history has shown that it is more a question of if, not when such power will be unleashed. Indeed, there is a cliché that suggests everything looks like a nail when you have a hammer. What, then, is to be said of the appearance of everything when you have an atomic bomb? Prophetic or no, Virilio's observations here, and the observations like it throughout his work, bring into view aspects of his work that tend to be underacknowledged by scholars who engage with his work on technology: when Virilio writes about war, he is often waging war on the idolatry of technology.

The next sections of the final portion of this chapter will dig further into the ways Virilio exposes the effects technological idolatry can have on the environment through its increasing of speed (*drómos*), harnessing of motion (*kīnēsis*), and manipulating of light (*phōs*). However, as a part of that work, it is also important to show how Virilio's identification, exposure, and dismantling of technological idols relates to and complements Ellul's work on technology as it has been described so far.

Fight Like Jacob

The image of Jacob wresting with God is one of the most biblically based examples that captures the connections between Ellul and Virilio's similar identification, exposure, and dismantling of technological idolatry. In numerous texts and interviews, Virilio describes the image's influence on his work in the following way, and it is worth quoting at length given its significance for this and other chapters:

> Here we are in a domain which is wonderful, but only provided that we fight against it. It's Jacob's wrestling match with the Angel. We must not lie down before the machine, we have to fight. I have had a photo of Delacroix's painting at Saint-Sulpice for years in my office [. . .]. It's extraordinary. This is an anti-idolatry fight. I think we haven't sufficiently analyzed what this episode meant. Because those themes from the Bible are unsurpassable: Babel, the Flood, Jacob wrestling the Angel, I could go on and on. What is Jacob? When he was beaten by the Angel, he was told, "You will no longer be called Jacob, but Israel." So Abraham, Isaac, Jacob, they're the invention of Monotheism (Jacob is the origin of Israel). That's not nothing. It's history. We rediscover a way to focus. What does Jacob do? If anybody is clever, Jacob is. What does he do? He fights. But he doesn't get himself in a fight to the death. He is in a fight for life. He wants to remain human before the God whom he recognizes as such. This is the very image of the opposition to idolatry. He is an emblem of anti-idolatry. This is the reason why Delacroix's painting is so extraordinary: Jacob has laid aside his bow, his spear, and his shield. His combat is hand-to-hand against the idol. He

> wants to worship God, but Jacob wants to worship as a human being, not as a servant. What I am saying is that here is the big question: we are in the process of raising idols. In this sense, we're lost. There's no future in it. If this theme has any sense whatsoever, if idolatry really is a major theme for us, then clearly idolatry has no future. As for myself, I am waiting on those Jacobs who will wrestle with the machines.[69]

Not only does this passage and the many others like it in Virilio's work affirm with a biblical frame what has been suggested so far about Virilio's fight with technological idolatry, it also identifies connections between Ellul's work and his own. Virilio never directly mentions Ellul as the one who brought this image into view in a way that would define his work on technology, but it is noteworthy that Ellul issues the following call in the conclusion to one of his most popular works, *Presence in the Modern World.* There, Ellul writes:

> We cannot make this genuine statement: "I planted, and God gave the increase." We cannot relinquish a situation into God's hands ("God will open eyes, ears, and hearts") until we have struggled continuously, to the point of daybreak, as Jacob did, until we have waged battle up to the limit of our strength and known the despair of failure, without which this so-called confidence in God and this orthodoxy are only hypocrisy, cowardice, and laziness. All I have written has no value unless it is understood as a call to arms, exposing which adversary much be confronted, which weapons must be used. Then, at the centre of this combat, the Word can be spoken—but not otherwise. When we have truly taken seriously the concrete situation of the men and women of our day, when we have heard their cry of anguish and understood why they have no desire for our disembodied gospel, when we have participated in their physical and spiritual suffering, in their despair and hopelessness, when we have entered into solidarity with our fellow citizens and our universal church, as Moses and Jeremiah did with their people, as Jesus did with the wandering crowds, sheep without a shepherd—then our voice will be able to proclaim the Word of God. But not before![70]

It is doubly noteworthy that Karl Barth invokes the same image and uses it to make a similar call in "The Problem of Ethics Today," a chapter in his early text *The Word of God and the Word of Man.*[71] Barth put the image to use in his work differently from Ellul and Virilio, but the image shaped Ellul and Virilio's work on technology in a similar way. It is likely, then, that Virilio heard (read) and responded to Ellul's call with a similar Jacobian fight against technological idolatry. At the very least, if Virilio's engagement with this biblical scene was inspired by someone else or for another reason, it still places Ellul and Virilio in the same biblical territory for their combat with technology. This means, as Virilio puts it in this passage, that both thinkers were similarly engaged in an "anti-idolatry" fight with technology, and the image of Jacob wrestling with God was a significant inspiration for the fight. In terms of method, Ellul and Virilio differ. But they fought the same war. Ellul's method was sociology and history reinforced with Protestant Reformed theology and ethics. Virilio's was philosophy and phenomenology infused with Catholic theology and ethics. They were unified by a strong sense of what the modern world had to wrestle with: the damage caused by its construction, sacralization, and justification of technological idols. And this is why both thinkers opted for interdisciplinary approaches in their work. Essentially, they were using every method at their disposal to fight idolatry.

With this in view, we can now turn to examples showing Virilio engaging with Ellul's thought and bringing it into new territory. It is the first of many instances that show Ellul and Virilio similarly fighting like Jacob.

Speed, Motion, and Light

In the final chapter of *Open Sky*, Virilio engages with Ellul's *Humiliation of the Word*, citing his reading of God's speaking of light into existence in Genesis 1.[72] Ellul's interpretation of light's protological appearance is taken to be the aspect of creation, which unveils reality and reveals time. Then, Virilio references Ellul's fear of how modern technologies can confuse reality and truth through manipulations of discourse, a fear at the core of *Humiliation of the Word*'s thesis. Virilio proposes that, as uses of technologies powerfully increase speed (*drómos*) and enable more powerful harnessing of motion (*kīnēsis*) and manipulation of light (*phōs*), confusions of reality and truth have surpassed the boundaries of language and reached the point of ubiquitous confusions of time and space. Virilio concludes, then, that an addition to the Tower of Babel has been built at the twentieth century's end: a Babelian confusion of people's senses of time and space. He ends *Open Sky* with the cryptic suggestion that the turn of the century signifies a glimpse of the Tower of Babel's shadow; an ominous image of a vertical and dark pillar without any horizontal crossing; the shadow of a cross with its arms lopped off.[73] Consequently, Virilio asks, in response to Ellul's sociological analyses of technologies as they have been described in this chapter: How can people speak clearly about thought, action, and making in the world if people cannot even sense clearly the environments in which they are situated? Accordingly, Virilio focuses Ellul's work on *technique*, *technology*, and *propaganda* onto people's senses and the environments in which all technologies appear. In this way, he agrees with Ellul: *yes, technologies are distorting speech!* But he also adds to Ellul's analysis: *now, senses and the environment are being disrupted!*

The term Virilio frequently uses to describe his method for analyzing *technique*, *technology*, and *propaganda* is *dromology*, which he often says is derived from the Greek word *drómos*, meaning *race*.[74] *Race* is the word's most concise definition, but *drómos* also can be defined as *running*, *road, course*, *racetrack*, or *a ceremonial walkway to a tomb or a temple*. The term's multiple significations, when combined with Virilio's analyses of the effects technologies can have on the environment (*phusis*), motion (*kīnēsis*), and light (*phōs*), is hermeneutically important. It adds theological, especially Neoplatonic and scholastic, meaning to Virilio's *dromological* studies of technologies. This is another aspect of Virilio's work that typically is underacknowledged.

Neil Turnbull is the only scholar who identifies and elaborates on the influence of Neoplatonism and scholasticism on Virilio's writing. Turnbull argues, "We need today to reflect with a more heightened seriousness on the canonical status of Virilio's work [. . .] . Virilio is much more than simply a minor post-structuralist thinker, because—and unlike many of his purported post-structuralist brethren—he begins [. . .] the process of relocating modern philosophy on a mystical theological-train."[75] I agree with Turnbull. However, his related argument, "The key to understanding Virilio's work resides in a Neoplatonic metaphysics of light rather than a social theoretics of speed," misses how light and speed connect in Virilio's work.[76] By forcing an either/or reading of Virilio's emphases on speed and light then, Turnbull overcommits to the argument, "The key thinker 'behind' the Virilian oeuvre is Nicholas of Cusa (Cusanus)."[77] Cusanus certainly is an influence for Virilio, but there are other theological influences who are just as important in and beyond Virilio's focus on light and speed. Ellul, is one, of course. But there are others, and many of them show how his analyses of speed, motion, and light often are used to expose technological idolatry.

Bernard of Clairvaux is one of many theologians Virilio uses to expose how technological idols problematically, if not dangerously, alter the environment (*phusis*) as they increase speed, (*drómos*), harness motion (*kīnēsis*), and manipulate light (*phōs*). As Virilio writes in *University of Disaster*, " 'Light is the name for the shadow of the living light,' observed Bernard de Clairvaux. Bearing witness to this light born of the light of the 'Completely Other,' the *theoscopic* visionary also paid involuntary tribute to the photosensitivity of the historical fact."[78] Clairvaux observed, "The very ancient exposure of the image of the Universe that reaches us, after a bit of a time lag, thanks to the shadow cast by light-years that enable us to contemplate not only the stars, those 'maternity wards of humanity,' but the very origins of the galaxies, in the revelatory bath of cosmic history."[79] For this reason, "If the first screen is the Moon (the night star), the very first imagery of the Universe is the speed of the film of light that enables us to see the history of the past as composed of constellations—and does so without forgetting the cosmetic *morphing* that clouds the perception offered by the lenses of our famous telescopes. I'm talking about gravitational mirages and other black holes that distort the reality of the starry firmament."[80] Essentially, Virilio uses Clairvaux to capture with philosophy-phenomenology how technologies can alter our senses with manipulations of speed, motion, and light. In this particular scene, Virilio identifies the telescope as a technology that "morphs"—mediates and manipulates—our perception of the world. As it does this, something negative happens to our relationship with God, the Living Light, the Completely Other, and the shadow of this Completely Other Living Light: the reality of heaven, the starry firmament, and the God who made and governs it in love becomes distorted and distant for those who gaze in distorted fascination at the world through the perception morphing lens of a cylindrical scope. All a potent metaphor for modernity's perception and environment morphing idolatry of technology.

Later chapters will draw out other theological influences and themes in Virilio's work to substantiate these critiques of Turnbull's otherwise insightful article. But here, it is important to keep the general character of Virilio's philosophical-phenomenological analyses of technologies in view. Taking the above into consideration, it is helpful to see Virilio as the kind of thinker who shows more than he says what *technique*, *technology*, and *propaganda* have become in the modern world. For instance, in *Art of the Motor*, Virilio quickly moves from the Industrial Revolution to the twentieth century in a way that Ellul would not, writing, "After the shipwrecks and derailments of maritime and rail acceleration, the collisions and crashes of the car and the plane and the steam press, and then the rotary press and the rotary image press of cinematic illusion, wave trains would come along and produce their own specific catastrophes through radio and video signals."[81] When this happens, "News is dynamite, information explodes like a bomb, opinion polls or war propaganda are time bombs."[82] The motifs of "shipwrecks" and "information" exploding "like a bomb" frequently are used in Virilio's writing, and they will be examined in detail later. But in terms of the complementary similarities and differences between Ellul and Virilio's work on technology, Virilio plays with words in ways that Ellul's typically clear, at times didactic sociological-historical analyses do not. Yet, despite Virilio's typically intricate, at times convoluted wordplay, his phenomenology shows precisely and concisely what *technique, technology*, and *propaganda* are individually, and what they can do together, without having to define them at length. Some good examples from his work for this have already been provided earlier, and there are more to come. However, in general, the similarities and differences in method and concentration described here demonstrate not only how Ellul and Virilio's work relates to each other, but also why the writings of Ellul and Virilio on technologies are stronger when read in conversation. Whereas, Ellul primarily examines how technologies, and the idolatry of technology, can change the way people speak and write about, and consequently enact, ethical thought, action, and making today, Virilio primarily examines how technological idolatry can change people's senses and the environments in which ethical thought, action, and making take place. By bringing the two thinkers into constructive conversation, it becomes clear that their critical analyses of technologies, with their respective emphases on language and justification, and senses and mediation, are important for theology and ethics today. For these reasons, the forthcoming chapters will continue to bring Ellul into conversation with Virilio.

Now, if all of this represents technology's idolized status in modernity for Ellul and Virilio, how did technology reach this state? For Ellul and Virilio, the first parts of the answer to this question are discovered in the Industrial Revolution and Greco-Roman antiquity, but only the first parts. The Industrial Revolution and Greco-Roman antiquity may show how modern technology came to be what it is today, but biblical texts like Genesis 3 and 4 reveal deeper truths such as why we produced technology in the first place and why we are prone to idolatry. Accordingly, the next chapter moves from Ellul's sociology-history and Virilio's philosophy-phenomenology to their theology and ethics of technological idolatry. To make this move, I use the hermeneutic of *myth* for Ellul and Virilio's work on technology to show how they think technology became what it is today, why humans produce technology, and why we are prone to idolizing it and other things that are the products of our thought (*epistēmē*), action (*poiēsis*), and making (*tékhnē*) in the world.

NOTES

1. Martin Heidegger, *The Question Concerning Technology*, trans. William Lovitt (New York: Harper Perennial, 2013), 8–16. When Ellul mentions Heidegger, he does so unfavorably and it usually is to suggest that his association with National Socialism delegitimizes all of his work. See, for example, Jacques Ellul, *Reason for Being*, trans. Joyce Main Hanks (Grand Rapids: Eerdmans, 1990), 148–149. Yet, Ellul does mention "dasein" and having read Heidegger's writing in Ellul, *Freedom, Love, and Power*, 32–33 and 74–75. Ellul also mentions "the situation of *Dasein*" as constitutive of the situation of prayer in Jacques Ellul, *Prayer and Modern Man*, trans. C. Edward Hopkin (New York: Seabury Press, 1973), 59.

2. Ellul and Troude-Chastenet, *Politics, Technology, and Christianity*, 7; 11; and 88 offers a timeline for the development of Ellul's work on *technique* as well as Ellul's description of the research and writing process for *Technological Society*. Ellul's lack of engagement with Greek philosophy in his sociology, theology, and ethics grows from a principled aversion to metaphysics and scholasticism rather than

ignorance of classic Greek philosophy. See Jacques Ellul, *To Will and to Do*, trans. C. Edward Hopkin (Philadelphia: Pilgrim Press, 1969), 16–20 for one of Ellul's many indictments of scholasticism. See Jacques Ellul, *Histoire des institutions: L'Antiquité* (Paris: PUF, 2016), 107 and 165 for a rare case in which Ellul summarizes Plato and Aristotle's thought.

3. For example, Jacques Ellul, *Subversion of Christianity*, trans. Geoffrey W. Bromiley (Eugene: Wipf and Stock, 2001), 22–27.

4. Jacques Ellul, *Technological System*, 128 and 325. See also Ellul, *Empire of Non-Sense*, 114 and 187; Jacques Ellul, *Living Faith*, trans. Peter Heinegg (San Francisco: Harper and Row, 1980), 262–263; Ellul, *In Season and Out of Season*, 28 and 51; Ellul, *Technological Society*, 187; Jacques Ellul, *Théologie et Technique*, eds. Yves Ellul and Frédéric Rognon (Genève: Labor et Fides, 2014), 142 and 238; and Ellul and Troude-Chastenet, *Politics, Technology, and Christianity*, 88. David Lovekin, *Technique, Discourse, and Consciousness* (Bethlehem: Leigh University Press, 1991), 70–71 also identifies the connection I am making here between *tékhnē* and *technique*.

5. Ellul, *Technological System*, 24.

6. Ellul, *Technological Society*, 76–79.

7. Ibid., 77–147 lists and defines these five characteristics. See Lovekin, *Technique, Discourse, and Consciousness*, especially 70–71 and 79–89, for a philosophical study of Ellul's *technique* in general and its connections to *tékhnē* in particular. Van Vleet, *Dialectical Theology*, 77–84 situates Ellul's work on technologies among other continental and analytic twentieth century critiques of technologies, whereas, 84–115 describes the "characteristics" and "consequences" of *technique*. Goddard, *Living the Word*, 134–137 and 143–146 also systematically summarizes important characteristics of Ellul's sociology of *technique*.

8. See David Lovekin, "Art and Technology: All or Nothing," *Empire of Non-Sense*, for a summary of Ellul's work on modern art and *technique*. I adapt Lovekin's references to Rembrandt's work on page 29 because they are helpful for this section's descriptions of *technique*.

9. Ellul, *Empire of Non-Sense*, 63–80 argues this at length, however, here I am summarizing Ellul's larger argument about modern art because it helps capture his basic understandings of modern *technique* and classic *tékhnē*—not because I want to start a debate about the relative value of modern art.

10. That being said, Ellul was not ignorant of and did appreciate some modern artists. For example, in the afterword to Ellul, *Empire of Non-Sense*, 167 he writes with irony, "Heinrich Böll, Nathanael West, Marc Chagall, Salvador Dalí, Luis Buñuel, Igmar Bergman, Federico Fellini, Saint-John Perse, Pablo Neruda, all have failed. Who will set sail on the high seas?" That is, each of their respective "failures" marked their successes. In Ellul and Troude-Chastenet, *Politics, Technique*, and *Christianity*, 136 he says, "Take Dali's painting in one instance[:] there is Christ which for me is one of the most beautiful crucifixion paintings ever. Then we have cubism as the other instance. These two examples show how art is split. The first represents the helplessness and suffering of man in the face of technology and second is a work that reflects the chilling perfection of the technological system." See also Ellul, *Autopsy*

of Revolution, 52 and 122 for similar, and favorable, references to Soviet filmmaker Sergei Eisenstein; and, Ellul, *Hope in Time of Abandonment*, 56–57 and Ellul, *Technological Bluff*, 388–389 for favorable references to Italian filmmaker Federico Fellini.

11. Ellul, *Empire of Non-Sense*, 467–48 and 65–66.

12. Ellul, *Technological Society*, 340.

13. See, for example, Jacques Ellul, *La Pensée Marxiste*, eds. Michel Hourcade, Jean-Pierre Jezequel, and Gérard Paul (Paris: La Table Ronde, 2018), 110.

14. Ellul, *Technological Society*, 340.

15. Ibid., 3.

16. Karl Marx and Fredrick Engels, *Marx and Engels Collected Works Volume 35: Capital Vol. I* (London: Lawrence and Wishart, 1996), 81–82.

17. Ibid., 83.

18. Ibid.

19. Ellul, *La Pensée Marxiste*, 49 and 69.

20. Ellul, *Technological Society*, 54–60 and 183–208.

21. Ellul, *Technological System*, 138–145; 270–272; and 300–304.

22. Ellul and Garrigou-Lagrange, *In Season Out of Season*, 175–176; Ellul, *Technological System*, 142; and Ellul, *Perspectives on Our Age*, 26.

23. Ellul, *Technological Bluff*, xv.

24. Ellul, *Technological Society*, 132. Ellul, *Théologie et Technique*, 45.

25. Ellul, *Technological Society*, xxv–xxvi; Ellul, *Technological System*, 22–33; and Ellul, *Technological Bluff*, xi–xvi.

26. Ellul, *Technological Society*, 79–80.

27. Ellul, *Technological Society*, 323–324.

28. Ibid.

29. Ibid. Also, because this aspect of Ellul's work is so important, see the following for other examples of what I am referencing here. Ellul, *Technological Society*, 47; 142; 175; and 367–368; Ellul, *Technological System*, 149; 152; and 281–282; Ellul, *Théologie et Technique*, 50–51; 268–269; and 284–285; and Ellul, *Technological Bluff*, xv–xvi; 31; and 216–217.

30. Ellul, "Ends and Means," *Presence in the Modern World*, 39–62 is noteworthy in this regard.

31. For example, Ellul, "An Ever More Perilous World," *Living Faith*, 217–234.

32. Ellul and Virilio use different translations of the Bible throughout their writings and their citations will be left as written. All direct citations of the Bible in this book are from the New Revised Standard Edition unless otherwise noted.

33. Ellul, *Technological Society*, 144–145.

34. Ibid.

35. Ellul and Troude-Chastenet, *Politics, Technology, and Christianity*, 63.

36. Jacques Ellul, *Humiliation of the Word*, trans. Joyce Main Hanks (Grand Rapids: Eerdmans, 1985), 22–23.

37. Ellul, *Propaganda*, x.

38. Ibid., 62.

39. Ibid., 79.

40. Ibid. See 62–70 for an extended comparison of political and sociological *propaganda* and 70–79 for a comparison of political agitation and social integration propaganda.

41. Ibid., 65.

42. Ibid., x.

43. Ibid., ix–xviii provides a good summary of modern *propaganda*, its difference from earlier forms, and why it is important to take the modern phenomenon of *propaganda* seriously.

44. Ibid., 113–114.

45. Ellul, *Technological Society*, 347.

46. Tony Judt, *Postwar*, 221.

47. Ibid.

48. Ibid.

49. Ibid. Of course, Coca-Cola is not the only example. It has been documented at length in Bernhard Rieger, *The People's Car: A Global History of the Volkswagen Beetle* (Cambridge: Harvard University Press, 2013), how Volkswagen, particularly the original Volkswagen Beetle, underwent a significant transformation in, through, and by *propaganda* and advertising before, during, and after World War II. The Beetle initially appeared in Nazi propaganda alongside the mythical Aryan human. Then it ascended to advertising prominence after the war with the now famous *Lemon* and *Think Small* advertisements. Similar to Coca-Cola in Europe, then, Volkswagen became a normal part of everyday life in North America even if what it initially represented was offensive and threatening.

50. Ellul, *Propaganda*, 148. Ellul's emphasis.

51. Ibid., 165.

52. Ellul, *Technological Society*, 144–145. The imagery of "power," "nakedness," and technology appears throughout Ellul's writings on technologies. For example, Ellul, *Humiliation of the Word*, 98–99; *Meaning of the City*, 99 and 166–167; and Ellul, *Living Faith*, 257–258. Later chapters examine the importance of this imagery for Ellul's theology and ethics.

53. James A. Hijaya, "The *Gita* of Robert Oppenheimer," *Proceedings of the American Philosophical Society* 144, no. 2 (June 2000): 123–167 for relevant references to this quote and account of the Bhagavad Gita's influence on Oppenheimer's philosophy.

54. Ellul, *Propaganda*, 146–147.

55. Virilio and Armitage, *Virilio Live*, 72.

56. Ibid.

57. Ibid.

58. Hirsch, "Complexity in the Age of Simplicity," 118.

59. Ibid.

60. Virilio and Lotringer, *The Accident of Art*, 87.

61. Paul Virilio, *A Landscape of Events*, trans. Julie Rose (Cambridge: MIT Press, 1996), 32–33.

62. Ibid.

63. Virilio, *Pure War*, 34.

64. Derek Hawkins, "Brian Williams is 'Guided by the Beauty of Our Weapons' in Syria Strikes," *Washington Post*, April 7, 2017, accessed November 27, 2020, https://www.washingtonpost.com/news/morning-mix/wp/2017/04/07/beautiful-brian-williams-says-of-syria-missile-strike-proceeds-to-quote-leonard-cohen/.

65. For example, Paul Virilio, *The Information Bomb*, trans. Chris Turner (London: Verso, 1998), 109–110; Paul Virilio, *Strategy of Deception*, trans. Chris Turner (London: Verso, 1999), 77; and, Virilio and Armitage, *Virilio Live*, 98.

66. Paul Virilio, *The University of* Disaster, trans. Julie Rose (Cambridge: Polity Press, 2007), 89–90.

67. Ibid., 119.

68. Virilio, *Negative Horizon*, trans. Michael Degener (New York: Continuum, 1984), 173. Virilio's emphasis.

69. Paul Virilio and Sylvère Lotringer, *The Accident of Art*, trans. Michael Taormina (New York: Semiotexte, 2005), 79–80. For other instances in Virilio's work where the same image appears, see: Paul Virilio, *The Art of The Motor*, trans. Julie Rose (Minneapolis: University of Minnesota Press, 1993), 84; Virilio and Armitage, *Virilio Live*, 157; and, Virilio and Lotringer, *Crepuscular Dawn*, 170.

70. Ellul, *Presence in the Modern World*, 93–94.

71. Karl Barth, "The Problem of Ethics Today," *The Word of God and the Word of Man*, trans. Douglas Horton (London: Hodder and Stoughton, 1928), 142–143.

72. Paul Virilio, *Open Sky*, trans. Julie Rose (London: Verso Books, 1995), 139–142. Virilio's reference is to the section "Light" in Ellul, *Humiliation of the Word*, 211–214.

73. All references in this paragraph are to Virilio, *Open Sky*, 139–145. Virilio uses the imagery of Babel similarly throughout his writing. See, for example, Virilio and Petit, *Politics of the Very Worst*, 79; Virilio, *A Landscape of Events*, 3–4; Virilio, *Information Bomb*, 71 and 107; Virilio, *Strategy of Deception*, 48; Virilio and Lotringer, *Crepuscular Dawn*, 34 and 60; Paul Virilio, *City of Panic*, trans. Julie Rose (Oxford: BERG, 2004), 43; Virilio, *University of Disaster*, 19 and 41; Virilio, *Grey Ecology*, 63; Paul Virilio, *The Futurism of the Instant*, trans. Julie Rose (Cambridge: Polity Press, 2009), 52–53; Virilio and Brausch, *A Winter's Journey*, 113–114; and, Virilio and Richard, *Administration of Fear*, 29–30.

74. For extended examples, see Paul Virilio, *Negative Horizon*, 101–118 and 129–132. For shorter examples, *Politics of the Very Worst*, 14 and 57–58; Virilio and Armitage, *Virilio Live*, 71; Paul Virilio, *The Original Accident*, trans. Julie Rose (Cambridge: Polity Press, 2005), 91–92; Virilio, *Grey Ecology*, 27; Virilio, *University of Disaster*, 19; 87; 123; Virilio and Richard, *Administration of Fear*, 15–16; Paul Virilio, *The Great Accelerator*, trans. Julie Rose (Cambridge: Polity Press, 2010), 64.

75. Turnbull, "Light and Illumination," 261.

76. Ibid., 260.

77. Ibid., 262.

78. Virilio, *University of Disaster*, 73.

79. Ibid.

80. Ibid.

81. Virilio, *Art of the Motor*, 24.

82. Ibid. For other examples of Virilio's phenomenological descriptions of *technique*, *technology*, and *propaganda*, see Paul Virilio, *War and Cinema*, 53 and 67; Paul Virilio, *The Vision Machine*, trans. Julie Rose (Bloomington: BFI Publishing, 1988), 11; Ellul and Lotringer, *Pure War*, 33; Virilio, *Landscape of Events*, 29–30; Virilio, *The Information Bomb*, 28–29 and 62; Virilio and Armitage, *Virilio Live*, 122–127; Paul Virilio, *Ground Zero*, trans. Chris Turner (London: Verso, 2002), 15 and 24; Virilio, *Desert Screen*, trans. Michael Degener (London: Continuum, 1982), 17; Virilio, *City of Panic*, 30–31; Virilio and Richard, *Administration of Fear*, 38–50; Virilio, *The Great Accelerator*, 64; Paul Virilio, *Art and Fear*, 74.

Chapter 3

Combatting Myth with Myth

The previous chapter focused primarily on what Ellul and Virilio think technology has become in the modern world: a powerful and dangerous idol that should be wrestled with, just as Jacob wrestled with an angel of God. It also highlighted some of the methods Ellul and Virilio use to identify, expose, and dismantle an idol, which, for both thinkers, is modernity's quintessential idol. The methods highlighted were sociology-history for Ellul and philosophy-phenomenology for Virilio. As a conceptual precursor to the next chapter, this chapter outlines how *myth* functions in Ellul and Virilio's work on technology in order to provide the theoretical basis required to see how both thinkers use *myth* to bring revelation (theology) into their histories of technology, and as they do this, use theology to attack modernity's idolization of technology.

SOCIOLOGICAL MYTH

New Demons offers Ellul's most elaborate and thoroughgoing commentary on his understanding of *myth* in general, and sociological *myth* in particular. Ellul also refers to and uses theological *myth* in his work, but he does so in a less forthcoming and extended way. The latter kind of *myth* will be featured in the next section because, as I will show, sociological and theological *myth* function dialectically in Ellul's work on technology and the two need to be separated conceptually in thesis/antithesis fashion before they are brought together in the clash he creates between revelation and history.

Ellul thinks sociological *myth* powerfully institutes difficult to define orders of sacred and profane in societies, and through such institutions, offers individuals justification(s) for existence. Explaining these connections,

Ellul writes the following in *New Demons*: "If [myth] is a grasp on and a justification for man's situation *hic et nunc*," and if *myth* "is an image of the most mysterious depths of man in confrontation with a given reality, then [myth] cannot, by its very nature, be the same now as then."[1] Still, *myth* is a constant historical phenomenon. It "necessarily appears in specific forms, but its characteristics and reasons are constant and common to all."[2] They are constant and common because sociological *myth* sits at "the nerve centre of [the] natural and social structure" to knit a society together and provide people with justifications for existence—including justifications for thought (*epistēmē*), action (*poiēsis*), and making (*tékhnē*) in the world.[3] Sociological *myth* is able to do this, Ellul says, because it gives societies a semblance of substance to difficult to define orders of pure/impure, holy/blemished, cohesion/dissolution, respect/violation, permitted/forbidden, life/death, and, most importantly, sacred/profane.[4] And these institutions, for many individuals, become bases for perceptions and articulations of, essentially justifications for, existence. Another way of putting this is that sociological *myths* establish a society's foundations, upon which their idols of choice are built, sacralized, and justified.

According to Ellul, sociological *myth* is so powerful that if a society did not have them at its nerve centers to institute sacred and profane orders, it would be difficult, if not impossible for that society function.[5] This is because pure materialism, he frequently points out, is intolerable for most individuals and societies because both individuals and societies need something or someone more than just material to establish the kinds of meaning and purpose people need in order to live, that is think, act, and make in the world. Sociological *myth*, then, is what constitutes that *something* or *someone* more, that transcendent something or someone, beyond the material that can provide meaning, unity, and purpose. In other words, however constructive or destructive a society's *myths* may be, and whether every individual in the society adheres to those *myths*, all societies require sociological *myths*, senses of the immanent and transcendent, in order to exist in a cohesive way.[6] Otherwise, that society will descend into chaos.

A country's flag, with its correlative *myths* and instituted sacred and profane orders, is one of many examples of the binding and loosing power of sociological *myth*. Something as arbitrary as colored fabric waving in the wind can bring a sentimental tear to an eye or kick off a violent rage, all because the sociological *myth* the flag symbolizes institutes a formidable, intuitively known and believed in sacred/profane order.[7] In some cases, all the flag and its *myths* do is cause citizens to shed tears and place hands upon their hearts. In other cases, the flag and its *myths* incite violence and war. But whatever the case, Ellul argues that *technique*, *technology*, and *propaganda* often are used to leverage sociological *myths* in ways that bind societies together, form them, and inspire particular thoughts, beliefs, and actions.[8] As the previous chapter highlighted, Ellul experienced this firsthand at a Nazi rally. After this experience, he repeatedly encountered similar social phenomena before, during, and after the war. All of these experiences stacked up and theoretically and practically showed Ellul how dangerous *technique*, *technology*, and *propaganda* can be when they are used to inflate sociological *myths*. Essentially, when *technique*, *technology*, *propaganda* use sociological *myths* as means to the end of societal engineering, they can manufacture pseudo (or actual) religious beliefs (idolatry) that cause people to think, do, and make things that they would otherwise consider abhorrent. The Nationalist Socialists are an obvious example, but Ellul says most modern nations play the *propaganda* and *myth* game too. Even the ostensibly democratic nations.[9]

As the previous chapter cautioned, however, the technicians leveraging sociological *myth* to form societies and engineer peoples' thoughts, beliefs, and actions are not solely to be blamed for this phenomenon. Such technicians do not, because they cannot, create sociological *myths* ex nihilo. Rather, if they want their *propaganda* to be effective, it can only manipulate or weaponize preexisting *myths*. Otherwise, the technician will have no kindling for the fire he or she wants to start. Consequently, societies tend to welcome *propaganda*, love *propaganda* in fact, because it provides certainty, purpose, and safety in a world that can often seem uncertain, purposeless, and unsafe. For this reason, *propaganda*'s stirring up of sociological *myths* is a "remedy for loneliness" in a society filled with lonely people.[10] The threat is that this coupling of *propaganda* and *myth* can "turn a normal feeling of patriotism into a raging nationalism [because it] not only reflects myths and presuppositions, it hardens them, sharpens them with the power of shock and action."[11] Again, Ellul's wartime experiences alerted him to this truth.

Just as it is the case with idols, the most effective kind of *propaganda* is the kind that society welcomes without really knowing what they are welcoming. The most powerful sociological *myths* are the kind that people in a society intuitively know and adamantly believe, but cannot fully articulate what they believe when they are called upon to do so. And, Ellul is able to show how, when *propaganda* and sociological *myths* reinforce each other, they can be incredibly powerful and dangerous. The terms "belief-image" and "power-image" aptly capture the nuanced layers of sociological *myth* as they are described in *New Demons*, as well as other texts, because sociological *myth* essentially produces the belief and power-images that institute the sacred in any given society. In turn, *propaganda* manipulates these belief and power-images to form societies and provoke particular thoughts, beliefs, and actions.[12] But what forms do the most powerful sociological *myths* assume in the modern world and what are their instituted sacred orders?[13] Ellul argues that history and science are the West's two fundamental *myths* with technology and the nation-state their instituted scared orders.[14] His analysis of "the four poles" of the modern sacred is important and of interest to Ellul scholarship, but for this chapter and book as a whole, it suffices to focus on the hermeneutic significance of modern *myth* and its instituted sacred orders as it relates to Ellul's work on technology. With that in view, Ellul writes, because Western "civilization is atheistic (not areligious, but simply not recognizing any formal deity to be worshipped as such)," sociological, "myth[s] today will not wear the mask of any active gods."[15] However, functionally, sociological *myth* "contains an element of belief, of religious belonging, of the irrational, without which it could never express what it is meant to express for man."[16] For this reason, "It cannot be said that man is no longer religious just because Christianity is no longer the religion of the masses. To the contrary, [modern man] is just as religious as medieval man."[17] The only difference is that the West's religion, or its sense of what is sacred and profane, has changed. When Christendom declined in the West, new sociological *myths* and sacred orders replaced its old *myths* and sacred orders. And in a strange reversal, what caused the decline of the West's old *myths* and sacred orders—history, science, technology, and the nation-state—introduced into the West new *myths* and conceptions of the sacred.[18] It is a common reversal that happens in history, Ellul says. What or who initially desacralizes the sacred and demythologizes the *myths* of a society often ends up becoming the new sacred and *myth* of that society. In this sense, the sacred's presence in history is quite similar to the spirit Jesus speaks of parabolically in Matthew 12:43-35.

It is important to note that Ellul's account of sociological *myth* and its institution of sacred orders, both in relation to and beyond technology, does not mean he resents Christendom declined and Christianity lost its status as a motivating belief and power-image in the West. Ellul actually welcomes the decline, because he sees it as an opportunity for Christians to (re)discover their iconoclastic, desacralizing, and demythologizing roots. For this reason, he writes, "[Today] we are faced with a situation to which we are not accustomed and for which we are scarcely prepared. This is a good, and we can thank God for the liquidation of the era of Constantine and for the end to the monumental error which was Christendom."[19] Now, "To the very degree to which the Church and Christians are witnesses to the one true God who has revealed himself, they have the stern duty of desacralization and demystification to perform in society."[20] This duty requires, "A 'No' on the religion which man constructs on the basis of his works. It is directed toward man's sacralizing—his calling sacred—his own works, the myths he builds up [. . .] to magnify himself and his works."[21] The church is called to do this, Ellul believes, because "we are witnesses to the true God, we have to reject such sacralizing, just as the first Christians rejected the calling sacred of nature, and the Reformation likewise rejected the sacred progressively restored in the Middle Ages. We have that operation to resume in the present age."[22] The next section will show how Ellul performs what he describes here as a "duty of desacralization and demystification" in society. But in the final parts of this section, I need to establish why Ellul believes the church's duty to desacralize modernity's idols and demythologize the sociological *myths*, which lay the foundations for such idols, is so important.

Ellul believes that if the church allows sociological *myths*, with their institutions of sacred and profane orders, to retain their power, then danger is on the horizon. Examples have already been provided in this section in respect to war and violence in general. However, more particularly, and in regard to technology, modern sociological *myth*, and the sacred, Ellul writes the following in *New Demons*—and what he writes is taken word-for-word from *Technological Society*. As it was quoted in the previous chapter from the latter text, Ellul writes in the former text that an indication of "of man's senses of the sacred in technology is the care he takes to treat it with familiarity. It is well known that laughter and humor are frequently a person's reaction in the presence of the sacred. That is true of primitive peoples, but it is also the reason why the first A-bomb was called 'Gilda,'" and so on.[23] Today, he observes, "It is fine to possess the power of the atom," but "[for man] to find himself all alone with [atomic power] in his hands, to know that he is responsible for his decisions, with only his strength to count on that is an unbearable situation."[24] Now, the last chapter described how Ellul exposes the technician ascribing sacred significance to the bomb in order to make the thought of atomic annihilation bearable. So it is unnecessary to take up that line of inquiry again here. Instead, it is important to inquire here about the broader society in which this technician exists, to reflect on what it is about the technician's society that inclines he or she to sacralize the bomb that they built. Putting this in the form of a question, how does the technician and the society in which he or she exists learn to ascribe sacred significance to technological objects like these and become comfortable with such dangerous forms of sacralization (idolatry)? Sociological *myth*, combined with technique, *technology*, and *propaganda* is Ellul's answer. A primary way that a society can justify making let alone using a technology of this kind is the formation and adherence to particular *myths* that delineate what and who is sacred, and what and who is profane. In this case, the society in view believes the *myths* that permit it to say, and feel itself justified in saying, *our society and our bombs are sacred and their society and their bombs are profane*. Yet, the enemy society being profaned does the same thing and this causes the hand of the atomic clock to move closer to midnight. Such sacralizing and profanation by sociological *myths* explain why 56 percent of Americans still think dropping atomic bombs on Hiroshima and Nagasaki was justified.[25]

With the above in view, the church's duty of desacralization and demythologization is not a game for Ellul. It is, as he sees it, a grave responsibility because it holds the church and the world back from the construction, sacralization, and justification of idols that can cause significant violence, if not total annihilation. To show how Ellul performs this desacralization and demythologization of modern *myths* and its sacralizing of technology, the next section focuses on Ellul's understanding and use of theological *myth* in his work.

THEOLOGICAL MYTH

Ellul's theological redescription of *myth* takes it as a form of revelation that appears in history and cannot be described fully or directly. His theological account of *myth* demythologizes sociological *myths* and desacralizes their institution of sacred and profane orders. This gives Ellul conceptual grounds to write about technology's evolution from a dialectical, rather than a singular perspective, and it helps him create a confrontation between revelation and history.

Discussion of theological *myth* appears in Ellul's writing, but less so than his discussions of sociological *myth*. For example, in *Meaning of the City*, he writes in a footnote as he comments on the Tower of Babel narrative:

> I hesitate to use the term "myth," for at best it is obscure, appearing in many diverse forms. From the conscious myth of the Greeks to Sorel's conception, from the biblical myth to the "twentieth-century myth," from the legendary myths of the Indo-Europeans to the Bultmannian, there are vast differences in form, meaning, and value. Barth's criticism of any conception of myth in the Bible is in truth valid only for a relatively minor part of the different possible meanings of the word. He is attacking the mode of interpretation in the Bible used by the historical school, also by form criticism, and finally by Bultmann.[26]

Ellul cautiously uses the word *myth* because people tend think the word refers to fictional, fanciful stories. But as he clarifies here, *myths* vary from culture to culture, and not all *myths* should be defined as such. Whenever the word *myth* is used, he suggests, we ought to ask: What kind of *myth* are we talking about? Answering this question, Ellul says, gets us closer to an understanding of what the *myth* in view means for the culture in which it is formed. Once this is established, the *myth* can reveal important truths about the culture in which it exists. After making this point, Ellul offers his own definition of theological *myth*:

> When I use the word [myth] I mean this: the addition of a theological significance to a fact which in itself, as a historical (or supposed to be such), psychological,

> or human fact, has no such obvious significance. Its role is therefore to make a fact "meaningful," to show it up as bearing the revelation of God, whereas in its materiality, it is neither meaningful nor the nature of revelation. This is how myth operates. It does not destroy the historical reality of the event, but on the contrary gives its full dimensions. In [the Tower of Babel] myth, God assigns a name to a people who would like to make a name for themselves and quite naturally assigns a name taking its meaning from God's act, from the confusion of languages, but also a name with meaning for the whole story, which is one long confusion—the confusion between man's power and God's.[27]

In other words, when Ellul uses the term *myth* when engaging with scripture and theology, he is not describing something he thinks did not happen. Rather, he is describing an event in which history is confronted by God's revelation and given meaning beyond itself. The Tower of Babel is the example he uses here. As I have written elsewhere about Ellul's understanding of Genesis 1–3, he would answer affirmatively if the question *did it happen?* was put to him about the Tower of Babel story. He would also add that it probably happened in a way that "calls into question everything solid and taken for granted."[28] This is because, for Ellul, theological *myth*, along with all of God's revelation, "shakes the foundations of the most elaborate cultural edifice; it stirs up uncertainty. It seizes hold of everything that constitutes our humanity, our state and societies, our hierarchy of values, our morality and our religion, and puts it all in the crucible."[29] Ellul is influenced by Barth here, because both thinkers believe God's self-revelation is unpredictable, untamable, and explosive; existentially intimate at times, but ultimately impossible for a creature to grasp, let alone elaborate fully.[30] Ellul is also influenced by Kierkegaard, in that he believes God's truth is often communicated via indirect communication, and in contrast, that direct communication or full recognition of God's truth, is in most, if not all cases, a mark of paganism.[31] Altogether, this means Ellul thinks theological *myth* reveals truths in and beyond the realities that make up our past, present, and future.[32] And, when theological *myth* comes into contact with history sociological *myth*, revelation starts a fight with history.

Ellul's deployment of these Barthian and Kierkegaardian themes in his use of theological *myth* is critical for his dialectical writings, and especially so for his dialectical writings on technology. As noted in the previous section, Ellul states early on in his work that the church has a duty to demythologize sociological *myths* and desacralize its sacred and profane institutions, including the idols that *myths* and their sacred orders incline people to construct, sacralize, and justify.[33] Now it can be shown how Ellul performs this demythologization and desacralization. "One cannot at the human level," Ellul writes in *False Presence of the Kingdom*, "Combat a myth except through myth."[34] In Ellul's case, he combats sociological *myth* with theological *myth* through a dialectical confrontation that can be described in the following way: on the one hand, there are the powerful justificatory sociological *myths* and instituted sacred orders, which the church is called to demythologize and desacralize as part of their witness to God (the sociological-historical side of Ellul's work); on the other hand, there are the theological myths, which demythologize and desacralize the powerful sociological myths and their instituted sacred orders (the theological-ethical side of Ellul's work); and, when the two sides of this dialectic are brought into contact, sparks fly.[35]

With these definitions and use of sociological and theological *myth* in view, it becomes clear why Ellul criticizes Rudolf Bultmann and Paul Tillich's respective approaches to *myth*. As Ellul writes in *New Demons*, "Such cultivated intellectuals as Bultmann and Tillich bluntly adopt [the] monolithic position [that] since modern man is imbued with science and no longer believes past legends and myths, since his motivations are rational, since he reasons and is absorbed in techniques, [then] he is rational and has left the mythical mentality behind him."[36] Not so, Ellul responds to Bultmann and Tillich. *Myths* have not been completely abandoned in the modern world. Only the old *myths* have been abandoned and new *myths* have appeared. In fact, contra Bultmann and Tillich, Ellul contends that the sociological *myths* of science and history have become the new *myths* of the modern world.Then, a society's belief in these *myths* and their institutions of the sacred can only be described as religious (idolatrous).[37] So, if the modern world now looks to history and science for its understanding of the sacred, and it is technology and the nation-state that are identified as the new sacred by these *myths*, then the church and its theology should not capitulate to, but combat these *myths* and their sacred orders. For this reason, whereas Bultmann demythologizes scripture, Ellul demythologizes society; whereas Tillich identifies correlations between theology and society, Ellul identifies counter-correlations.[38] In both cases, Ellul assumes this posture toward society, scripture, and theology because he believes the church should hold the modern world back from the destruction caused when its *myths*, sacred orders, and idols are unchecked and reinforced rather than exposed and resisted.

Theological *myth* is a critical aspect of Ellul's resistance to modern *myths*, the sacred, and idols, then, because it helps him bring revelation's explosive force to bear upon the sociological *myths* that exert pressure and generate power in the modern world. Ellul's use of the Tower of Babel narrative in the passage quoted earlier is an example. In this case, he uses the Babel story to shed light on the "confusion between man's power and God's."[39] As a document of something that happened in history, the Babel account does contain meaning in and of itself. However, the story also reveals meaning beyond the historical events. This meaning is revealed by a God who chooses to intervene when a society takes up any project to make a name for itself. From this perspective, this makes the Tower of Babel more than a tower that was built at a certain point in history. It makes the tower a symbol or a figure for any human project that confuses humanity's power with God's power. The atomic bomb, as one of many examples, is one such human project that can be viewed as an iteration of the Tower of Babel Story. Which is to ask: What does the Tower of Babel story reveal about the atomic bomb? It reveals that nothing good will happen when a nation tries to split open the sky in order to gain power over other nations. In this way, the Babel story reveals truths about and resists the kinds of sociological *myths* that incline us to justify the building and sacralization of monstrous technologies that open the sky in order to bring heaven—or hell—down to earth.

Finally, how societies perceive their origins are especially important aspects of the confrontation Ellul sets up between sociological and theological *myths*. Because, if you know the primary way(s) a society views its origins, then you can expose the roots of its current sociological *myths* and attack those roots with theological *myths* that offer a contrasting account of our origins. In Ellul's work, origins are often referred to as "returns to zero," because they represent the ground upon which all *myths* are built.[40] For this reason, returns to zero are not always constitutive of a society's actual origins. In fact, they most often constitute how a society perceives its origins, what a society views as the events that most define who they are, what their purpose is, and how this identity and purpose will define their thought, belief, and action in the present and future. As far as technology and modernity are concerned, Ellul writes the following in *New Demons*, and it features material taken word-for-word from an article he published twenty years earlier:

> The origin which [modern] myths are telling us about is no longer the same origin, nor is the event which they are interpreting the same. It is no longer the origin of the world and of man, because that is not a real question for people today. It is no longer the origin of the gods, for the traditional gods are indeed dead. It is no longer the advent of fire, or of the city. The origin, the advent, which enchants and at the same time obsesses [people], is the machine, electricity, the dominion over nature, abundance. If myth is always a return to zero, it certainly can be said that it is not always the same zero. In the western world today, our zero is in the neighborhood of the year 1780, in that marvelous time when all the hidden forces of nature were about to be let loose by a sort of magic, to be placed in the service of man. The myths of work, of progress, of history are constantly telling us how that happened. They are constantly causing us to relive this innovation and to participate in this dawn. This takes the place of the "why," and of all justification. At the same time they are showing us that it was truly an origin and not a fulfillment.[41]

There are a number of origins, or returns to zero, mentioned in this passage that are put into confrontation throughout Ellul's work on technology. For readers who are familiar with Ellul's corpus, the origin of the city will stand out as an allusion to his book *Meaning of the City*, his doctrine of the city, and his theological-ethical account of the origins of civilization and *technique*. That being said, there are other important origins mentioned here, especially the year 1780. This is a multilayered allusion to the Industrial Revolution (an obvious ground zero for modernity and its technologies); the French Revolution (regicide and the end of the *ancien régime* in 1789); and finally, the French Republican Calendar (a return to zero for the first year of the Republic, which, for *technical* reasons, began in 1792 rather than 1789). All of the latter historical moments mentioned represent the West's perception of critical moments in modernity and technology's origins. As such, they are the roots of the West's current sociological *myths*: science and history with technology and the nation-state as their sacred domains. Conversely, the former origins mentioned, the origins of the world, of humanity, and the city, represent returns to zero that no longer really register in modernity's consciousness. They are the *myths* that no longer retain power in the West, and consequently, they no longer designate the orders of sacred and profane in the modern world. However, these *myths*, these theological *myths*, still retain power and they can be used, as Ellul uses them, to confront, and demolish if he has his way, the sociological *myths* that hold power in the world today. In this sense, Ellul does not only engage in direct combat with the idol of technology. He also picks a fight with the *myths* that made it possible for technology to become idolized in the first place.

Later sections of this chapter will feature in detail the sociological and theological *myths* that confront each other in Ellul's work on technology—specifically, the Industrial Revolution, Greco-Roman Antiquity, the Cain and Abel narrative in Genesis 4, and the Fall account in Genesis 3. The next sections, however, will feature Virilio's complementary use of *myth* in his work on technology. For, even if Ellul and Virilio's uses of *myth* in their work on technology differ on the surface, they are remarkably similar at their core, and therefore, mutually reinforcing.

HISTORY AND REVELATION

It is essential to see how Virilio approaches revelation and history to see how he uses *myth* in his work on technology. But history as a concept and as material for study constitutes a significant portion of Virilio's work. So, similar to my engagement with Ellul's work with history, revelation, and *myth* in the previous two sections, this or the next section does not cover the entirety of Virilio's work in these areas. Instead, it focuses only on Virilio's understanding of the relationship between history *and* revelation to provide context for the next section that examines the role of *myth* in his work on technology. With that said, I do hope that what follows will inspire readers to (re)visit Virilio's work on history with a fuller understanding of this aspect of his work. To that end, broader claims will be made at certain points here to provoke (re)appraisals of Virilio's work on and with history. But, to reiterate, this section's chief aim is to describe Virilio's view of revelation and history's relationship, because it will constructively frame the next section, which focuses on *myth* in his work.

The opening chapter to *Landscape of Events* provides one of the concisest and clearest accounts of Virilio's understanding of revelation and history's relationship. He opens with a description of God emphasizing particular aspects of the many divine attributes present in traditional Catholic conceptions of God. According to *Landscape of Events*, and other key texts that thicken its thinner theological content, Virilio thinks God is the eternal, "completely other,"[42] creator who exists apart from and is omnipresent to and in creation.[43] As Virilio puts it,

> For God, history is *A landscape of events*. For Him nothing really follows sequentially since everything is co-present. From the smallest fact to the greatest historical event, "There is nothing wonderful before Him," [Ecclesiasticus 39:20]. This transhistoric land, so hard to imagine, extends from age to age, from eternity to eternity. There—though where exactly?—generations have been "pouring from times of endless date," in their ever-changing nature—"in their going, in their flowing"—outlined against the horizon of an eternal present. We can hardly hope to grasp this atemporal perspective, in which before and after coexist, unless we see it as a film. Only, a film in which the sequence shot would constantly keep the beginning and the end in view[.] A landscape of time in which events suddenly take the place of relief, of vegetation; in which the past and the future loom up together in all their obvious simultaneity.[44]

In other words, history, along with all of creation, is bracketed and enveloped by God's eternal omnipresence and omnipotence.[45] Her, Virilio offers two images to help time-and-space-bound creatures such as ourselves to envision God's eternal gazing upon, and potent omnipresence to creation—essentially, what we creatures experience as time, space, and history and what God experiences as creation. The first image, that of a film depicting its beginning and ending in at all times as the story of the film plays out, is as a generally scholastic and particularly Thomist account of creation. It is a vision of creation in which everyone and everything created by God is set in motion at its creative beginning and intended and given the capacity to move toward its end ordained by God. The second image, however, strikes me as a theological image quite unique to Virilio. It depicts God perceiving and interacting with what we creatures experience as historical events growing from the ground of history (i.e., creation and created time) like sculptures and plants.

However, Virilio also emphasizes throughout his work that, just as many of events of history will amount to beautiful sculptures and plants, there also are many that will not.[46] Fortunately, he also shows that it is not entirely up to we creatures to discern on our own which events are destined for beauty and for terror. As we see in the passage earlier, Virilio thinks God's vision of history as a landscape of events is to some extent available to we creatures if/when God makes this vision knowable through revelation—otherwise, how would Virilio be able to describe this vision of God if he did not encounter this vision in some way or form? Which is to say, the only way Virilio is able to share this vision of God with readers is if this vision of God has been shared with Virilio. And Virilio has seen and shares a vision, or visions, wherein history's events are incomplete sculptures and partially grown plants. Consequently, an implicit eschatology emerges in his work, one in which we time-and-space-bound creatures cannot see all the monuments and vegetation springing up from the landscape of history, and we cannot fully know how God will perceive these sculptures and monuments in the end (the *eschaton*). Yet, similar to Paul in 1 Corinthians 13:12, Virilio believes we can, because of who God is and what God does, see in part—before we see in full—what these sculptures and plants are—and will be—as events bloom around us from history's soil.[47] Or, less poetically, it could be said that Virilio's understanding of history takes Heidegger's view of history, being, and time, and fills it up with strong Scholastic and apocalyptic motifs. That is to say, whereas Heidegger conceptualizes history as a flowing stream of experience, which we experience and apprehend in our localized times and spaces, Virilio (1) conceptualizes history as a film reel, in which all that happens in the present moment is bracketed and enveloped by God's gaze and the presence of God and (2) conceptualizes history as soil, from which revelatory events pop up from the ground like monuments and vegetation.[48]

With the aforementioned context in view, the light imagery Virilio uses throughout his work hints at God's omnipresent and omnipotent bracketing and enveloping of history, and in turn, offers readers a good frame for his understanding of revelation and history. Which is to say, Virilio thinks God's light can illuminate history's darkness, making it so that its monuments and vegetation are not destined to remain in complete darkness, half/malformation and partial/zero growth forever.[49] To provide a few of many possible examples, Virilio's frequent use of Ephesians 5:13—"All that appears appears in light"[50]—and Bernard of Clairvaux—"Light is the name of the shadow of the living light"[51]—captures how his work in general, and his work on history in particular, is brightened by the eternally omnipresent, omnipotent, and omniscient light of God. Sometimes this illumination is explicit, other times it is implicit, and there are moments in Virilio's work where the light of God flares brightly and beautifully. But, as even casual readers of Virilio are aware, light imagery flashes most in his work when he is forming phonological descriptions of past, present, and future wars, atrocities, violence, and accidents. So why, then, is Virilio apparently more preoccupied with the past, present, and future's darkness—even when light is in view—than he is with God's revelation in history? Or, more simply, if Virilio has experienced powerful revelations of God's light, why does he spend so much time focusing on the dark sports of human history? How one answers these questions is critical for Virilio scholarship. They determine whether one reads Virilio's references to scripture, theology, and God as quite inconsequential or quite consequential for a significant portion of his work.

If it is not obvious by this point, I will put forward directly here that I am inclined to view these facets of Virilio's work as quite consequential for much of his work. With this stated outright, the first part of the answer to these questions, to which I will return at a later point, is that when Virilio focuses on the dark spots of history, he is often doing a form of negative theology, which, when read with this hermeneutic in mind, can bring readers into contact with the brilliant light of God that he mentions directly and positively in only certain, unique places in his work. The second part of the answer, which I will elaborate on here, emerges with a look at Virilio's use of Walter Benjamin's *Theses on the Philosophy of History*, particularly his reworking of the famous essay's engagement with Paul Klee's *Angelus Novus* (Angel of History). Returning to *Landscape of Events*, Virilio writes in the opening chapter: "Walter Benjamin once wrote, 'A Klee pairing named *Angelus Novus* shows an angel *looking as though he about to move away from something*. His eyes are staring, his mouth is open, *his face is turned toward the past*. Where we perceive a chain of events, *he sees a single catastrophe*. It keeps piling wreckage upon wreckage and hurls it at his feet.' "[52] Then, he writes the following:

> Today this theological vision no longer belongs to the angel of history. It belongs to each and every one of us. From now on, the acceleration of the reality of time causes revulsion at the being-here-present. Like fright, which results in the body's retreat, evaporation of hope in the future causes us to regress, inducing permanent resentment. The work that follows [in this book] is devoted to a radical reversal in perspective. It is halfway between an essay and a narrative in which a series of atypical events succeed each other over the course of only a dozen or so years; events whose scope has escaped the theorists as much as the historians of the moment.[53]

It is important that Virilio says Benjamin's use of Klee's *Angelus Novus* is a *theological* rather than a *philosophical* vision. He is correct in describing it as such insofar as, beyond using the image of an angel, Benjamin brings Messianism and Messianic time into the essay. For example, Benjamin writes of "a conception of the present as the 'time of now' being shot through with chips of Messianic time" and "every second of time [being] the straight gate through which the Messiah might enter."[54] This creates a tense, paradoxical image of despair and hope in respect to the angel's vision of history. On the one hand, there is despair: the angel sees history as a big catastrophe, a heap of wreckage, and big clump of empty, homogenous time. On the other hand, there is hope: there are bits of Messianic time in the catastrophe and wreckage that the angel can see, and each of these Messianic chips are surging with possibly of a Messiah appearing to halt the catastrophe, redeem the wreckage, and fill up the time of human history until it overflows with meaning, power, and peace that transcends the scope of history's material catastrophe.[55] As seen in this passage, Virilio's critical addition to Benjamin's image of history is the suggestion that this singular vision of the angel is now made available to every person in the modern world. How does Virilio think this has become possible?

Although the answer to this question is marginal to the discussion at hand, it is central to the focus of this book, so it is worth answering partially here. Virilio thinks the angel's vision is now made available to most if not all people in the modern world primarily through the development and expansion of technologies that have become characteristic of our historical moment. In this context, this is why Virilio's says that such technologies are "the best" because they provide many opportunities for us to discover and contemplate the material of history; and why also says, paradoxically, that they are "the worst" because they can manipulate or even explode the material of history, cause more catastrophes, and provoke us to turn away from history altogether.[56] Here, Virilio's conceptualization of "the tyranny of real time"—a singular experience of time and space that connects everyone in the world, and in the process, replaces, and often obliterates, all localized experiences of time and space—helpfully encapsulates his treatment of this phenomenon and the threats he sees on the horizon due to the angel of history's vision becoming ubiquitous in a technologically globalized world.[57]

Moving back to the primary discussion at hand, though, the ubiquitous *Angelus Novus* phenomenon brought about by modern technologies is at once threatening and promising for Virilio. This is because it forces upon us, in way perhaps unique to modernity, a "duty to remember" history to the extent of contemplating it intently, intensely, and honestly.[58] This is for two reasons. First, Virilio thinks we have a duty to contemplate history because, in general and for all people, if we do not remember our history, then we increase the likelihood for repetitions, if not amplifications of its evils.[59] Second, and more particularly for Christians, the Messiah has entered time in the form of God's Son, Jesus Christ, and this powerful, past, present, and future transforming self-revelation of God introduces an apocalyptic dimension into history that God's people are meant to search for and contemplate with intention, intensity, and honesty. As Virilio puts it in an interview, "The Bible is the book of the Apocalypse. It means that the Apocalypse is happening all the time, everyday, since Genesis. It never stops."[60] After the Old Testament comes, "The New Testament, you have Christ, and then you have the Apocalypse, *Alles fertig* [*it is finished*, a reference to Christ's cry from the cross in John 19:28–30]. History is following this same course, there is no doubt about it."[61] So, a unique kind of general/special revelation approach to history materializes in Virilio's work wherein history is (1) generally revelatory for all people because it is crisis inducing and instructive and (2) particularly revelatory for some people because they have encountered and been transformed by the redemptive and hope-bestowing presence of God. But either form of revelation will not be discovered if we do not remember, and constantly and critically practice, our duty to remember.

To clarify before moving forward, Virilio does not use the actual terms *general* and *special revelation* in his work. However, the way these concepts operate in his work becomes clear when, for example, he uses "chronos" and "kairos" to describe modern concepts and experiences of time, often in his discussions of technology.[62] Which is to say, for Virilio, the quantitative, often homogenous, empty, and sequential technologically enforced time regimes of *chronos* in the modern world constitute the time and space of general revelation; whereas, the qualitative, unique, full, and potential-filled time of *kairos* constitutes the time and space of special revelation in modernity's historical moment, and all historical moments. Accordingly, for the people of God, both kinds of revelation and time—*chronos* and *kairos*—are freely given, but only if they do not forget who God is and how God claims and forms a people who have a particular history, with a particular identity, and therefore a particular way of being in the world.[63] And of course, a common problem for the people of God in both the Old and New Testaments is forgetting what they would do well to remember: who their God is, and what their God has done in the past, and consequently will do in the present and future.[64] That is why, though it may not be evident to most readers, Virilio's believes the church is essential for the people of God and the world. The church not only *reminds* the people of God—and hopefully the world—about who God is and what God has done and does in the past, present, and future; the church also *mediates* in and for the world who God is and what God does in the past, present, and future.[65] As Virilio puts it, with resonances of James 1:23–27 in the pop culture reference, "The preacher in John Huston's film, *Wise Blood*, expresses [a truth] pretty clearly: 'The church of Christ without Christ is nothing but your own shadow, nothing but your reflection in a mirror.' "[66] Putting Virilio's passing theological remark in fuller theological terms, he believes, and he shows in his work, how Christ's presence in the church can move creaturely focus away from the shadows and mirrors of our time-and-space-bound existence and turn them toward God's light, which can illumine our past, present, and future. But, as he warns in this quote, if Christ is not really present in the church, then we time-and-space-bound creatures will only be gazing at shadows and reflections of ourselves and our history filled with violence. Then, more violence and less peace will abound.

The aforementioned context explains why Virilio's writing often assumes what could be described as the perspective of Klee and Benjamin's angel of history: it takes on this perspective to focus the reader's attention on apocalyptic events in history. At least, this perspective affects an instructive crisis for readers; at most, it causes an instructive apocalyptic event for readers, which flowers into a revelation of God's presence. So, Virilio is not being trite when he quotes Psalm 35, saying: "Apropos the false wit of these erudite demiurges who can tell so much just by looking at the Earth and the sky, but are, on the other hand, quite incapable of telling the time we're living in, let's turn to the Psalmist again: 'Truth shall spring out of the earth; and righteousness shall look down from heaven.' "[67] Virilio sincerely believes that truth will spring out of the earth because of God's self-revelation, which is known must fully in and through the incarnation, life, death, resurrection, ascension, and imminent return of Jesus Christ. Further, Virilio believes it is possible for all people to encounter this revelation in history as the monuments and plants of the landscape of events are contemplated intently and intensely. This, I think, is why Virilio is known for saying such things as "revolution is over; we're entering the era of *revelation*"[68] and "I am not a revolutionary; I am a *revelationary*."[69] I also think this is why, as I will show in more detail in the next chapter, he says "negativity is a positive task"[70] and that "the accident is an inverted miracle, a secular miracle, a revelation."[71] As he pushes the reader to contemplate the negative monuments and vegetation of the landscape of events, he is trying to create apocalyptic, crisis inducing, and revelatory moments for the reader, which cause the truth of God to spring up from the ground upon which the reader and all of humanity stands.[72]

With all of this said, it is important to specify that even if it were possible for Virilio to force a revelation of God upon and for the reader, he does not attempt to do so. Instead, he writes only enough about history's moments, monuments, and people to provoke the reader to research what he is referring to, contemplate the history in view, and then hopefully experience a true revelation as they do so. Then, at critical moments, Virilio will help readers experience this form of revelation by giving them the scriptural and theological references and motifs to illumine the past, present, and future material in view. Accordingly, if Ellul's approach to history and revelation tends to be expository, Virilio's approach to history and revelation tends to be descriptive. This preference for description over exposition, Virilio says, was the influence of his teacher, Maurice Merleau-Ponty. As he says in an interview, Merleau-Ponty's, "Research was directed towards the unreflected element in our perception of the world and the description, not the explanation, of the real. He kept open the dialectic of history."[73] Like Merleau-Ponty, Virilio focuses the reader's perception on that which they do not tend to reflect upon, he describes more than he explains, and he leaves open the dialectic of history. This is because he wants to create time and space for the Messiah to enter into the reader's localized time and space; to create time and space for the reader to be mystified by what is perceived in the past, present, and future; and, once they are mystified, encounter the revealed Messiah who can—and for Virilio, did and does—enter into the time and space of creation.

In contrast, Ellul for the most part leaves the dialectic of history open, but he does so in a different way. Except for a few of his texts, Ellul separates his sociology-history from his theology-ethics, thereby requiring the reader to engage the thesis/antithesis dialectic he sets up, and, if they want to, determine how to read his sociology-history into his theology-ethics and vice versa. But then in the 1970s, Ellul closes the dialectic of history with an explicit turn to universal salvation (i.e., total redemption and recapitulation) of history with the return of Jesus Christ.[74] Although most readers of Ellul take his turn to universalism and closing of the dialectic to be sincere, there are is a small contingent of Ellul readers who think this turn, and the closing it causes, is an attempt at Kierkegaardian irony. Whether or not such an attempt at Kierkegaardian irony is successful or constructive is debatable. However, either way, the possible ironic turn can be described in the following way: by closing the dialectic of history with a turn to universalism—the proclamation that everything and everyone in history will be saved and redeemed—Ellul attempts to paradoxically reopen the dialectic of history that has been closed by a modern world filled with intellectuals led toward closure by the death of God movement and dogmatic Marxist materialism. But all of this represents somewhat of a digression. The primary point to make here is that both Virilio and Ellul generally leave the dialectic of history open in their work so as to create time and space for a powerful revelation of God, but without making their attempts become forced or spiritless—until the 1970s, when Ellul is either sincere or insincere in his closure of the dialectic of history.

For now, a relatively simple way to summarize Virilio's view of history—along with the comparison to Heidegger's view of history, being, and time as provided above—is to take it as similar to Michel Foucault's "archaeology" method, but charged with an operative understanding of, and sincere belief in the revelatory, crisis inducing, redemptive, and hope-bestowing presence of the Messiah.[75] Or, better yet, Virilio's apocalyptic archaeology of history is like an *apocalyptic architecture* of history because he often insists that he is more an "architect" and an "urbanist" than a "philosopher" and he often alludes or directly refers to Augustine's *City of God* when he does so.[76] Accordingly, read with this in mind, Virilio emerges as an archaeologist-architect who surveys the city of man in order to uncover evidence of the city of God's insurgent convergence with and conversion of the city of man. Or, to return to Benjamin's influence on Virilio's view of history and revelation: imagine Benjamin's concept of history, but one in which the hoped-for Messiah has entered, and continues to enter, through the door of time into humanity's past, present, and future.

Of course, as beautiful as Virilio's apocalyptic approaches to history can be, they are also terrifying, just as any truly apocalyptic event is. For, as Virilio puts it, the violence of modernity's past, present, and future, so often caused by its powerful technologies, unveils to us the shocking truth that "the absolute death of nuclear arms, the possibility of suicide open not only to an individual, a society, or even a civilization, but to an entire species, reintroduces the question of God—the question of ethics on another level."[77] It is a level in which, "God has come back into history through. The door of terror."[78] In other words, Virilio believes God's apocalyptic entry into time and space is certain, but we cannot be certain whether this entry will inspire hope or terror.

With the aforementioned context in view, I am comfortable suggesting, both in relationship to passages like the ones cited above and the others highlighted in this section, that much of Virilio's work on and with history can be constructively read as a form of negative theology in which the times and spaces of history that seem to reveal God's absence actually, when contemplated intently and intensely with the aid of the church's presence and witness in the world, reveal God's presence—with *Landscape of Events, War and Cinema*, *Bunker Archaeology*, *Negative Horizon*, and *City of Panic* providing examples of this form of negative theology. As far as *myth* is concerned, the next section shows how Virilio's works with *myth*—usually the events, people, and structures of history that are of mythical status and proportion—often constitute "crosspoint[s]" between the most devastating material events, people, and structures of history and the most brilliant revelatory, redemptive presence and action of the Eternal and Divine One in history.[79] And with that said, the next section turns to *myth* in Virilio's work on technology.

MYTHS AS TENDENCIES

Virilio does not define *myth* as comprehensively as Ellul because he shows more than he says what *myth* is in the modern world. Still, a text like *Pure*

War, which features Virilio's most extended commentary on this aspect of his work, shows similarities in both thinker's approaches to *myth*. Just as they do for Ellul, *myths* give Virilio critical insight into the structures and functions that form modernity. But, whereas Ellul's "four poles of the sacred" suggests history and science are modernity's fundamental *myths*, with technology and the nation-state as their instituted sacred orders, Virilio uses Georges Dumézil's theory of the three structures (the sacred, the military, and the economic) and their corresponding functions (the priest, the warrior, and the peasant) to expose the *myths* and senses of the sacred that he thinks most drive the modern West.[80] On the one hand, Virilio would grant history, science, technology, and the nation-state shape the modern West in significant ways, to the extent of mythologization and sacralization. On the other hand, he would argue the military structure and the warrior function pervade these four poles, and consequently, the most powerful *myth* of modernity is war, with technology is its sacred engine. As Virilio puts it in *Pure War*, "My relation to the war-machine has always had a mythic dimension."[81] "I don't deny the importance of the return of the sacred because, in reality, it was caused by the nuclear age. The nuclear age has put us closer to the apocalypse."[82] With this in view, this section will show how the mythic dimension to Virilio's study of the war machine appeared during his decades-long study of the bunkers, blockhouses, anti-aircraft shelters, and submarine bases abandoned after World War II; whereas, his sense of the modern war machine's sacralized engine was awoken by the construction and detonation of atomic bombs.

Virilio acknowledges Dumézil's theory of the three structures and functions as an influence on his understanding and use of *myth*, but he also reworks the theory in novel ways to reinforce his studies of modernity's tendential movement toward violence, death, and war. In *Pure War*, Virilio describes the influence as follows: "Myths have an analytical capacity that cannot be denied. By comparison sociology seems a surface effect. What interests me is the tendency."[83] He continues: "Myth is tendency. [Dumézil's] three functions seem to me analyzers of the knowledge of war, of political knowledge, achieving infinitely more than all the successive macro- or micro-developments [of sociology]."[84] While Ellul often is referred to as a *sociologist*, he would not disagree with Virilio's appraisal of modern sociology. Actually, Ellul expresses similar reservations concerning studies of society that focus too much on sequential, macro- and micro-phenomena. Though they get at it in different ways, what both thinkers are trying to get at in their analyses of the society in which they are embedded, are the meso-structures, layers, developments, institutions, and other such phenomena that are resistant to the erosive effects of time, but also evolving as time passes, and consequently most determinative for the society they study. This is, of course, not to suggest that either thinker ignores macro- or micro-phenomena in their work. Rather, it is to highlight their similar critiques of modern sociological method. Both think modern sociology can become too fixated on linear studies of surface or granular phenomena, and that *myth*, by comparison, offers unique analytical insight into past and present social phenomena. For this reason, they tend to avoid bird's eye or microscopic views of the society they are in, and as far as *myths* and the sacred are concerned, they often look at the *stories*, which most define their society; how those stories (i.e., *myths*) delineate its sense of sacred; and therefore, form and drive it in difficult to articulate, but nevertheless real, ways. And, finally, if a society's primary *myths* delineate its primary senses and expressions of the sacred, then a society's primary senses and expressions of the sacred shed light on its primary idols. In this way, *myth* is the body of water that feeds the stream of a society's idols.

As for Dumézil, *myth*, and the sacred's influence on Virilio's work, Dumézil thought there is a "tripartite social structure" present in the origins and development of Indo-European society: the sacred, the military, and the economic. To this tripartite, social structure corresponds the "three functions" that shaped Europe as it developed, and to a certain degree, shapes modern Europe today. Those social functions, or one could say *roles* or *types*, are the priest, the warrior, and the peasant.[85] Each of these structures and types retain their formative power in modern European society through *myths*, and as Virilio and Lotringer discuss in *Pure War*, they exert formative pressure through "mythical projections" onto, in, and through various social phenomena.[86] Now, Virilio does not offer much commentary on Dumézil's treatment of *myth* beyond claiming that "*myth* is tendency," by which he means studying *myths*—in this case, *myths* concerning the sacred/priest, military/warrior, and economic/peasant—can expose the tendential development of a society and its people. But, fortunately, Ellul does mention Dumézil, along with other prominent contemporaneous intellectuals who studied *myth*, in *New Demons* offering more insight into how someone like Virilio understands and uses *myth* in his work. Ellul writes:

> The analysis of the myths [has] led to a much deeper understanding of a certain ongoing quality in man, a certain relation of man with the universe, a certain structure of soul. Research was carried out in various directions, but it all came together in a central group composed of Jung, Callois, Eliade, and Dumézil. At the same time it was seen that these myths fulfilled diverse functions, and that one could, for example, distinguish between explanatory, etiological myths (whose purpose was to throw light on a place or a people, or on the origin of a custom or an institution), and ontological myths (which explained some profound, permanent reality of man and which displayed man's reflections about himself). Along this line, it would appear that there was perhaps no other means

of expressing those reflections, that from his remote beginnings, man had discovered a special language which alone was suited to his greatest depths and to a direct expression of the inexpressible.[87]

Given what Virilio says in *Pure War*, it seems his understanding of *myth* aligns well with Ellul's descriptions of *myth* in this passage. Like Ellul, he sees *myth* as being etiological and ontological because he thinks study of *myths* can "throw light" on the places, people, customs, institutions, and origins of a society. This is because *myths* reveal the basic "tendenc[ies]" that manifest, form, and drive social phenomena below the macro-level and above the micro-level of a society.[88] In Virilio's case, though, he thinks the sacred/priest, military/warrior, and economic/peasant *myths* throw more light on the modern world's development than Ellul's four poles of the sacred. But irrespective of the *myths* Ellul and Virilio emphasize and study in their work, they agree that the modern world's *myths* tend to institute technology as one of its primary senses and expressions of the sacred; that, as the chief idol of modernity, sacralized technology sits at the center and margins of the modern world's most formative *myths*.

The critical move Virilio makes in respect to Dumézil's tripartite structure and functions is his suggestion that, in the modern world, the military structure and the warrior function have perverted the other two structures and functions.[89] As he says in *Pure War*, "We find priests, warriors, and peasants in contemporary society with overlays that cause the priest to be no longer simply a preacher," and "the warrior no longer simply a soldier, but also a military engineer, or the engineer of advanced techniques."[90] Then, "The peasant is also the proletarian, but the proletarian is also the professor, insofar as teaching has been discredited, etc."[91] As the next chapter will show, "perversion"[92] and "perverted priest"[93] imagery shape Virilio's work in critical ways. Although here it is pertinent to focus on how Virilio thinks the military structure and warrior function pervert the other two. As Virilio observes, "All of us are already civilian soldiers, without knowing it. And some of us know it. The great stroke of luck for the military class's terrorism is that no one recognizes it. People don't recognize the militarized part of their identity, of their consciousness."[94] But when Virilio speaks about the "military class," he does not mean what most people would think he means: generals, sergeants, foot-soldiers, and the like. Instead, the social class he describes is as follows: "What I call the 'military class' includes all those who reason within [a] technological logic."[95] Essentially, "The so-called 'technocrats' are [the] military class. They are the ones who consider rationality only in terms of its efficiency, whatever the horizon. The negative horizon's apocalyptic dimension doesn't strike them."[96] First, Ellul's work on *technique*, specifically its drive toward rationality, efficiency, and its absence of limits, is apparent here, though inflected with Virilio's characteristic focus on a militarized modernity. Second, apocalyptic motifs appear in this quote in negative, though illuminating forms. Such motifs are used to describe a global, technocratic, economy that functions in a military mode as it ignores, if it does not work actively against, a positive apocalyptic horizon (i.e., God's divine economy).[97] As Virilio describes this enduring, pervasive human economy—constituting a negative image of the Triune God's divine economy—elsewhere, citing Goethe, "One frequently hears it proclaimed that 'Art [i.e. art, artists, and *tékhnē*] cannot be immoral,' whereas what ought to be said is that *it cannot be illegal.* In losing any sacred character, it long since entered the baneful Goethean triangle of 'war, commerce, and piracy, the three in one, inseparable' (Faust II)."[98] Third, Virilio connects the ubiquity of militarism and militant technologies to a ubiquitous militarized consciousness, knowledge, rationality, and science (*epistēmē*) that pervades the modern world, forms its identity, and consequently perverts both Dumézil's three structures and functions and Ellul's four poles of the sacred. In fact, Virilio concludes his discussion of the military class with the following commentary:

> [The military class is] that kind of unbridled intelligence which gets its absence of limits from technology, from science. The war-machine is not only explosives, it's also communications, vectorization. It's essentially the speed of delivery. When Esso tells the French national train company: "We'll stop delivering containers, materials, gasoline, oil, refining products, unless you guarantee us trains with 4000-ton capacity running at an average of 100 km/hr"; when Esso threatens to make do with trucks, it's already war. Pure War, not the kind which is declared [. . .]. It's war operating in the sciences. It's everything that is already perverting the field of knowledge from one end to the other: everything that is aligning the different branches of knowledge in a perspective of the end.[99]

First, note the "branches of knowledge" imagery used in the final portion of the passage quoted earlier is important to note in respect to theology's place in Virilio's work on technology. As the next chapter will show, the branches of knowledge referred to here are an allusion to the Tree of Knowledge of Good and Evil—what, for Virilio, reveals the roots of modernity's militarized and militarizing tendencies. Second, and more on point with the discussion at hand, Virilio describes how the speed (*drómos*) of modern technology's delivery systems wears down, if it does not obliterate the difference between economic and military logistics (more generally structures and functions). In this case, Virilio uses a dispute between Esso and France's national train company to make the following point: in the modern world, the speed and power of technological innovation has made it so that multinational corporations and public transport systems literally and metaphorically go to war over questions of speed, logistics, delivery times, and so on. And as they go to war over such things, the consciousness and identity of those in the oil company and the national industry become militarized when strategy meetings at the boardroom table become functionally equivalent to a meeting about tactics in a military war room. Or, to take another, more everyday example, when the speed (*drómos*) of modern technology's delivery systems make a smart missile (an explosive projectile) functionally equivalent to a smartphone (an ultra-fast communication device), it does not take long for the consciousness and identity of the modern individual to be militarized.[100]

What this technological phenomenon does, then, is militarize modernity, and force it into a state of (un)conscious, (un)declared, but pure, perpetual war. For, in a kind of world where the possibility of atomic war and annihilation is determined by technological reaction times of nano- and milliseconds—essentially decisions with catastrophic implications being made at the speed of light—this terrible phenomenon, which all of us have somehow learned to live with, covers the world like a film and perverts all experiences of space and time, even if the perversion is on the periphery of our senses and consciousness.[101] As a significant part of this phenomenon, the sacred/priest and economic/peasant structures and functions are perverted by the military/warrior structure and function, thereby making the military/warrior structure and function modernity's mythologized, sacred economy, and the priest and the peasant, ultimately, a warrior in the ways that they think (*epistēmē*), act (*poiēsis*), and make (*tékhnē*) in the modern environment (*phusis*). Consequently, Virilio acknowledges that modernity's most determinative *myths* and senses of the sacred can include or allow space for all three structures and functions, but he also argues that the military/warrior structure and function pervades and powers the other two. He repeatedly warns readers that if we think modernity's militancy is contained to what we normally think of when we think about the so-called military class, then we need to rethink our understanding of modernity's militancy. A good place to start this rethinking would be modernity's militant *myths* and our sacralized, militarized technologies, because they reveal just how much the military structure and warrior function has pervaded the modern consciousness and identity.

But how did Virilio arrive at such a provocative insight into the *myths* and senses of the sacred that pervade modernity? As suggested at the beginning of this section, it took shape during his decades-long study of the bunkers, blockhouses, anti-aircraft shelters, and submarine bases that were left behind after World War II. It grew as he witnessed, from a terrified and contemplative distance, the construction and detonation of the first atomic bombs.[102] As Virilio shares in *Pure War*:

> People speak of the history of war, of battlefields, of deaths in the family, but no one speaks of the military space as the constitution of a space having its own characteristics. My work is located within this concept. I suddenly understood that war was a space in the geometrical sense, and even more than geometrical: crossing Europe from North to South, from the shelters of the German cities to the Siegfried Line, passing by the Maginot Line and the Atlantic Wall, makes you realize the breadth of Total War. By the same token you touch on the mythic dimension of a war spreading not only throughout Europe, bur all over the world. The objects, bunkers, blockhouses, anti-aircraft shelters, submarine bases, etc. are kinds of reference points or land marks to the totalitarian nature of war in space and myth.[103]

The detritus and derelict architecture of war that few people want to acknowledge, let alone contemplate and discuss, are for Virilio technological manifestations of the *myths* and the sacred, which most form and drive modernity's past, present, and future. If you look at these objects and structures long enough as Virilio did, you can see a general truth emerge, and perhaps encounter a special revelation: for some reason, one which typically escapes our ability to perceive and articulate, the general human tendency is toward violence, death, and war; not love, life, and peace; and, in the end, we need something or someone outside of ourselves to bend our tendencies toward the latter, away from the former. As Virilio writes in *Bunker Archaeology*, "Contemplating the half-buried mass of a bunker, with its clogged ventilators and the narrow slit for the observer, is like contemplating a mirror, the reflection of our power over death, the power of our mode of destruction, of the industry of war."[104] Yet, "The function of this very special structure is to assure survival, to be a shelter for man in a critical period, the place where he buries himself to subsist. If it thus belongs to the crypt that prefigured the resurrection, the bunker belongs too to the ark that saves."[105] For Virilio, even the most brutal, mythologized and sacralized, manifestations of humanity's tendencies toward violence and death can become times and spaces when and where there are revelations of life and resurrection. And in the same way that the waste and abandoned structures of World War II revealed these truths to Virilio, so too can the events, people, structures, and technologies of *all* wars throughout history. Essentially, once you have the hermeneutic of the military/warrior *myth* that pervade modernity, you can, as Virilio does, swoop over history like Benjamin and Klee's angel and see just how much the stories that past and present societies tell, the things that past and present societies worship, and the things past and present societies do and make, reveal a tendency toward violence and death that inspires a constant search for technologies that effect violence and death faster, more powerfully. Yet, this truth often registers in social and individual consciousnesses and identities in a mythical register because it defies literal explanation, and disorients and mystifies a person even when they try to think, speak, and write about history in literal, objective terms. Indeed, Virilio's accounts of history's enduing violence generates the following mystifying question: how does one articulate in literal terms the atrocities of history as they gaze intently and intensely at the brutal plants and monuments springing up from the landscape of history? For instance, to take a modern example, how does one speak in literal terms, and not speak in mystified terms, when they are describing a world held hostage by potential nuclear annihilation—a terrifying reality, which pervades, and subtly militarizes, our everyday existence? *Myth* is a way to articulate this reality, and get to the heart of this reality, so that we might get a better grasp of how it even became a reality in the first place. So, to combat *myth* with *myth*, Virilio writes about the basic human tendency toward violence as it appears throughout history in mythical, mystifying terms to confront readers with a real, staggering truth: the origin of humanity's ostensible civilization, along with its technologies is not peace, but war.[106] The mythologization of history and the remythologization of revelation is thus Virilio's way of getting the reader's attention and to respond to the human tendency toward violence and the enduring historical phenomenon of war.

On the positive side of such negativity, I suggested in the previous section that when Virilio writes of such terrible events, people, structures, and technologies, he is often writing about the "crosspoint[s]"[107] of death/life and despair/hope that are, lest we forget, embedded in a time and space in which "every second of time [is] the straight gate through which the Messiah might enter."[108] Read from this perspective, Virilio's constant focus on violence, death, and war in his work becomes much less a negative perspective produced by a permanent despair at the state of the modern world, and much more a form of negative apocalyptic theology, often composed in the key of *myth*. Virilio writes in such ways to instantiate a revelation of the truth that even the absolute worst structures and functions of militarism and warriors throughout history are conquered and transformed by the life, death, and resurrection of Jesus; that the perverting projection of the military structure and warrior function throughout history, and lodged in the modern identity and consciousness in the forms of *myths* and sacralized technologies, actually is "shot through with chips of Messianic time."[109] As such, he engages *myth* in his work to provoke the reader's modern, militarized consciousness and identity, and expose the interstices in the brutal monuments and plants on the event landscape that are letting the light of God's revelation shine through. To provide one of many examples of how this plays out in Virilio's work, a passage quoted in a previous chapter can be revisited, this time with a fuller picture of the intent behind Virilio's negative, mystifying accounts of history and technology. In *War and Cinema*, Virilio writes:

> In Western Europe and the Soviet Union, a new breed of military and revolutionary leader was beginning to have [a] charismatic effect on the masses[.] Real power was now shared between the logistics of weaponry and of sound and images, between war cabinets and propaganda departments[.] "Propaganda is my best weapon," declared Mussolini. And this is how Rosavita described the Duce: "Go, Caesar . . . ! Your task is fulfilled. From Caesar has come Benito Mussolini, strong and powerful as History has never before displayed; *his will partakes of the supernatural, the Divine, the miraculous, of Christ among men.*" Just as unexpectedly, when Hitler was asked in private who was his great historical model, his answer was not Bismarck, as one might have thought, but Moses. Perhaps it has not been properly understood that these miracle-working dictators no longer ruled but were themselves *directors.*[110]

The military structure and warrior function may mythologically project itself upon, in, and through the modern consciousness and identity, sacralized technology may be the power that pushes this projection farther and faster into the future, but the crucifixion and resurrection of Jesus Christ is the true crosspoint of history that moves in the opposite direction. This crosspoint of history reveals the light of resurrection after the darkness of death, even in the people and events of history that seem, on the surface, to indicate the absolute absence of God's incarnation, life, death, and resurrection—people and events such as those described in the passage just quoted. Read from this perspective, the mystifying ways a brutal dictator like Hitler and his cynical technicians such as Speer and Reifenstahl appears in Virilio's work, often represents negative theology crosspoints in Virilio's work; intersections of total depravity and death of humanity with the total holiness and life of God on the landscape of events, which the angel of history, and perhaps we, can see if we look closely enough at the brutal monuments and plants springing up from that landscape. The final chapter will elaborate further on this aspect of Virilio's work, but here, to conclude this section and prepare for the chapter to come, the following can be said about the similarities and differences in Ellul and Virilio's engagement with *myth* in their work on technology: Ellul demythologizes modernity's history and remythologizes scripture to expose and dismantle modernity's technological idolatry; Virilio mythologizes modernity's history and remythologizes scripture to do the same. Both thinkers' methods have their strengths and weaknesses, but their primary, unified aim is to identify, expose, and dismantle the idol of technology in the modern world—and scripture and theology are crucial to this enterprise. With this in view, the next chapter begins with how Ellul's demythologization of history and society and remythologization of scripture and theology play out in his work on technology.

NOTES

1. Ellul, *New Demons*, 91–92.
2. Ibid., 50–52.
3. Ibid.
4. Ibid., 56–57. For other characterizations of sociological *myth* see Ellul, *Presence in The Modern World*, 47–51 and 60–70 and Ellul, *Technological Bluff*, 137–138 and 385–388.
5. Similar descriptions of myth or the sacred also appear in early articles, such as, Jacques Ellul, "La sacré dans le monde moderne," *Le Semeur* no. 2 (1963): 24–36 and Jacques Ellul, "Le mythe de l'environnment," *Economies et Société: Cahiers de l'E.S.E.A.* 7, no. 9 (1973): 1539–1554.
6. Ellul, *New Demons*, 145–148 gives some good modern examples.
7. Ellul's critique of "commonplaces" is another example, but it has been adequately covered by other scholars such as Katherine C. Temple, *The Task of Jacques Ellul* (Toronto: McMaster University PhD Dissertation, 1976), 161–168 and Goddard, *Living the Word*, 120–122. Suffice it to say here, Jacques Ellul, *A Critique of the New Commonplaces*, trans. Helen Weaver (New York: Alfred A. Knopf, 1968), describes on page 9 the reasons for his critique of the commonplace as follows: "Every society produces its commonplaces, just as a living body produces wastes. The commonplaces are the excrement of society. But it is worth noting that the evidence left by those who have disappeared is rarely evidence of their nobility. When the bird has flown, when the nest has been abandoned and scattered, you know it was there only by the little pile of excrement you find. And the first traces left by man, except for his own bones and those of his victims, are weapons—axes, swords, arrows, and flints—as if all man had been able to leave behind him were the traces of his instincts of death and destruction." It is a seriously entertaining work but also deadly serious for Ellul, just as he is deadly serious when he writes about sociological *myth*. For Ellul, commonplaces are indicative of sociological *myth* because truisms and clichés are barometers for what societies tacitly assume and believe about morality/immorality, order/chaos, sacred/profane, and so forth. One such revealing commonplace for Ellul is "The machine is a neutral object and man is its master" (cf. Ellul, *Commonplaces*, 226–235).
8. Ellul, *New Demons*, 145–148.
9. See Ellul, *Propaganda*, 232–259 for an extended discussion on this point.
10. Ibid., 148.
11. Ibid., 40–41. See pages 33–61 for further discussion of how sociological *myth* is manipulated and weaponized by propagandists and *propaganda*.
12. Ellul, *New Demons*, 103–112 for descriptions of "belief-images" and 105–114 for descriptions of "power-images."
13. Other scholars have written about the importance of sociological *myth* and the sacred to Ellul's dialectical writings in general, and his dialectical writings on technologies in particular, but they do not take the hermeneutic of *myth* for Ellul's work in the direction I do. See Katherine Temple, *The Task of Jacques Ellul*, 73–91

and 171–185; Michael Burdett, *Eschatology and The Technological Future* (New York: Routledge, 2015), 144–145 and 155–160; Gregory Wagenfuhr, *Plundering Egypt* (Eugene: Cascade Books, 2016), 130–151; Darrell Fasching, *The Thought of Jacques Ellul* (Toronto: Edwin Mellen Press, 1981), 6–12 and 34–43; and Goddard, *Living The Word*, 122–127. I disagree with Fasching who argues in "The Sacred, the Secular, and the Holy," *Ellul Forum* no. 43 (April 2014), https://journals.wheaton.edu/index.php/ellul/article/download/414/16/ that, "*New Demons* [is] the Rosetta Stone of Ellul's authorship—for the first time bringing sociology and theology together in one book." *New Demons* is not the first book in which Ellul brings his sociology and theology together—*Presence in the Modern World* is. Or even, arguably, *Theological Foundation of Law*. Also, *New Demons* provides valuable insight into the hermeneutic of *myth* for Ellul's dialectical writings, but so do other texts, which calls into question Fasching's description of *New Demons as the* "Rosetta Stone" of Ellul's work.

14. Ellul, *New Demons*, 103–112 and 105–114. See Ellul, *Reason for Being*, 60–69 and 151–158 for further discussion of *myth*, history, science, and progress. Similar discussion of how "the two great fundamental myths on which all other myths rest are science and history" appears in Ellul, *Propaganda*, 39–40 indicating that this thesis was central to Ellul's work early on, since *Propaganda* was published in 1965. Just as Ellul lived in the West and did his thinking and writing in and for that context, I am doing the same. However, in doing so, I do not intend to exclude the East from my thinking and writing. I am only thinking and writing primarily in and for the context in which I find myself embedded.

15. Ibid., 92.

16. Ibid.

17. Ibid., 64–65.

18. Ibid., 58–59. Also, 61–62 and 67.

19. Jacques Ellul, *False Presence of the Kingdom*, trans. C. Edward Hopkin (New York: Seabury Press, 1963), 8. Although I would prefer to render the title of this text "*False Presence in The Modern World*," given its correlation to *Presence in The Modern World*, I have chosen to leave the title as it is for citational clarity.

20. Ibid., 202–203.

21. Ibid.

22. Ibid.

23. Ellul, *New Demons*, 73–74. Ellul, *Technological Society*, 144–145. See the section "Propaganda, Produced by Technique and Technology" in chapter 2 for the full quote and related commentary. Side-by-side comparison of the two texts shows only minor translation variances.

24. Ibid., 92.

25. Bruce Stoke, "70 Years after Hiroshima, Opinions Have Shifted on Use of Atomic Bomb," *Pew Research Centre*, August 4, 2015, accessed December 10, 2020, https://www.pewresearch.org/fact-tank/2015/08/04/70-years-after-hiroshima-opinions-have-shifted-on-use-of-atomic-bomb/. Immediately after the attack, 85 percent of Americans thought it was justified according to a Gallup poll. In 1991, a Detroit Free Press survey showed 63 percent of Americans think the attack was justified. The most recent percentage identified in a 2015 Pew Research poll is, as already noted, 56

percent. So there is a steady decline in approval. It seems to me that the reason for this change in perspective is America's increasing distance from the kinds of sociological *myths* and *propaganda* that fostered a sense of justification for attacking Japan in, what I would argue, was an unjustifiable way. The dropping of "Little Boy" and "Fat Man" on Hiroshima and Nagasaki were more for scientific and experimental reasons than military and strategic purposes. Likewise, the bombings were unnecessary in view of the fact that the Allied victory had already essentially been secured. For evidence of this, see Gar Alperovitz and Martin J. Sherwin, "Op-Ed: U.S. Leaders Knew We Didn't Have to Drop Atomic Bombs on Japan to Win the War. We Did It Anyway," *Los Angeles Times*, August 5, 2020, accessed December 10, 2020, https://www.latimes.com/opinion/story/2020-08-05/hiroshima-anniversary-japan-atomic-bombs.

26. Ellul, *Meaning of the City*, 17–18.

27. Ibid., 18. The significance of "naming" and "making a name" in Genesis later. Ellul also comments on theological *myth* in relation to Genesis 4 on pages 1–2 of *Meaning of the City*. See also Jacques Ellul, *Judgment of Jonah*, trans. Geoffrey W. Bromiley (Eugene: Wipf and Stock, 2011), 10–11; 17; 37; and 67 for other early, and similar, descriptions of theological *myth*.

28. Jacques Ellul, *Living Faith*, 144–145. See Michael Morelli, "What's in a Name? Jacques Ellul's Reading of Naming in Genesis 1–3," *Jacques Ellul and the The Bible: Towards a Hermeneutic of Freedom* (Eugene: Pickwick Publications, 2020), 82–95 for a description of how this applies to Ellul's reading of Genesis 1–3.

29. Ibid.

30. Ellul, "God for Me," *What I Believe*, trans. Geoffrey W. Bromiley (Grand Rapids: Eerdmans, 1989), 167–187 offers a concise rendering of this understanding of God and God's self-revelation.

31. Ellul, *Reason for Being*, 118–119 features Ellul's discussion and references to Kierkegaard on the subject on direct versus indirect communication in relation to God's truth.

32. See Ellul, *Humiliation of the Word*, 9–26 for a good, extended treatment of how he views the differences between reality/seeing/images and truth/hearing/words both in relation to and beyond God's revelation.

33. Ellul, *False Presence of the Kingdom*, 202–205. For a fuller example of Ellul's early writing on *myth*, the sacred, demythologization, and desacralization, see 200–209. Even earlier than *False Presence of the Kingdom*, which was published in 1968, Ellul says much the same in *Presence in the Modern World*, which was published in 1948. See 48–62 and 63–68 in the latter text for examples.

34. Ibid.

35. See the section "Strategic Style" in chapter 1 for further discussion of Ellul's dialectic, and how he uses the image of positive and negative poles and sparks to describe the relationship between his sociological-historical and theological-ethical work.

36. Ellul, *New Demons*, 43–45. He similarly critiques Bultmann in Ellul, *Subversion of Christianity*, 155–156.

37. Although Ellul critiques Bultmann in passages like these he does affirm aspects of his work. See Ellul, *Ethics of Freedom*, 156 in which Ellul references in

positive terms aspects of Bultmann's writing. Likewise, Ellul critiques Tillich but also quotes him at length in *Hope in Time of Abandonment*, pages 111–112, suggesting Tillich's passage is "beautiful."

38. John Wilkinson, "Introduction," *The Meaning of the City*, trans. Dennis Pardee (Eugene: Wipf and Stock, 2011), viii provides a good but brief description of the relationship between *Technological Society* and *Meaning of the City* and how myth functions in that relationship. Wilkinson also suggests in passing what I am suggesting here in respect to differences between Ellul's and Bultmann's work on myth. Fasching, *The Thought of Jacques Ellul*, 6–12 also features similar discussion.

39. Ellul, *Meaning of the City*, 17–18.

40. Ellul, *New Demons*, 113.

41. Ibid., 113. The original article is Jacques Ellul, "Mythes Modernes," *Diogène* 6, no. 23 (September 1958): 23–40. Ellul reworks some of the content from the 1958 piece for *New Demons*, and in the fourth section titled "Additional Observations" (pages 114–121), he adds a critique of "the enthusiasm of [current] intellectuals for utopia."

42. See Virilio, *City of Panic*, 129–130; Virilio, *Art as Far as the Eye Can See*, trans. Julie Rose (London: Bloomsbury, 2005), 106; and, Virilio, *University of Disaster*, 73 for examples, of Virilio describing or alluding to God as completely other.

43. Virilio, *Landscape of Events*, x.

44. Ibid. The endnote for the first quote in this passage includes the full verse of Ecclesiasticus 39:19–20: "The works of all flesh are before Him, and nothing can be hid from His eyes. He seeth from everlasting to everlasting; and there is nothing wonderful before Him." The endnote for the second quote cites Herman Melville, "A Canticle," in *Battle-Pieces and Aspects of the War* (1866; New York: Thomas Yoseloff, 1964), 128–129 as its source. See also Virilio and Armitage, *Virilio Live*, 144 where Virilio describes the landscape of events in the following way: "It refers to a timeless landscape which belongs to teleological vision of Deus, that is to say, a vision that freezes history, from its beginning to its end. You could say that it is like Marey's chrono-sculpture, using chronophotographic images of moments of the flight of a seagull. My interest in trajectory, in trajection, and trajectiles, underpins this conception of landscape." Étienne-Jules Marey was a photography and by extension cinema pioneer who made vivid, time-lapse-like images of people, animals, and objects in motion. As is Virilio's thought-provoking tendency, he points out that the camera Marey designed and used for such photography resembles that of a rapid fire rifle and coincided with innovation in repetitive fire weaponry. See Virilio, *Negative Horizon*, 63 for an example. Also, *War and Cinema* as a whole constitutes a significant case study of what I am only touching upon here.

45. See Paul Virilio, "Kinematic Optics," *Polar Inertia*, trans. Patrick Camiller (London: SAGE Publications, 1990), 37–53 for an extended treatment of God's divine attributes and how God relates to creation. Virilio blends this account of God's character and activity with commentary on modern physics, most notably Einstein's theory of general relativity as he does in other texts. Although the accuracy of Virilio's engagement with modern physics has been called into question by actual

physicists, I think chapters like these represent, at the very least, thought-provoking effort to form connections between well-known scientific theories and scholasticism, neoplatonism, and the like. Then again, I am not a physicist and cannot comment with any kind of authority on the legitimacy or lack thereof of regarding Virilio's attempts to blend theology, ethics, and physics in the way he does. See Alan Sokal and Jean Bricmont, "Paul Virilio," in *Fashionable Nonsense: Postmodern Intellectuals' Abuse of Science* (New York: Picador, 1998), 169–176 for a particularly harsh criticism of Virilio's work with physics deployed by two modern physicists.

46. Ibid., 14–15. Making this point in a chilling way, Virilio sets the following scene in *Landscape of Events*, writing, "Thirty years ago, the Empire Theater in Paris showed a very long newsreel, nearly ten hours' worth. If I remember rightly, its title was *Forty Years of History*." He goes on to describe how he and his friends hunkered down to watch the full ten hours of footage as they consumed sandwiches, beer, and coffee: "For a whole day, with short intermissions, we saw the first 'prewar,' the First World War, the interwar years, the Second World War, and finally *our own postwar* parading past." Then, he narrates the following revelation springing up from the soil of history's landscape of events: "We looked on all this as a traveling shot forty years long. As in a slow-motion film in which you see plants growing, buds developing, flowers unfolding, and petals falling, we could feel Hitler growing beneath the Kronprinz and the stratospheric rockets flying off beneath Georges Guynemer's plane. It wasn't history, not really a documentary. For us it was a sort of vision of the great circus of Time, of this *landscape of events* that God alone can contemplate." So, here, as in so many other places in Virilio's work, the connections between God's gazing upon and omnipresence to and in creation makes the stuff of history more than the stuff of history. It is a means by which God reveals Godself, and mediates revelation. In this case, the revelation is a guts-churning one in which Adolf Hitler, and the events surrounding him, emerges in and from history as a flower—in this case a *poisonous* plant—blooms from the soil. I think what Virilio is getting at here with such a negative revelation is that intent and intense contemplation of the past, present, and future may actually help us prevent poisonous plants from flowering in the soil of history.

47. Virilio, *Aesthetics of Disappearance*, 27. This passage will receive more attention in a later chapter.

48. It is, of course, beyond the scope of this section, chapter, or book to explore the similarities and differences between Virilio and Heidegger's views on history, being, and time. However, I consider this comparison worth making here in passing because doing so may offer a helpful way to understand Virilio's approach to history—at least for readers who are familiar with Heidegger's approaches to these things. Virilio is clear that he considers himself more akin to Husserl than Heidegger (cf. Virilio and Lotringer, *Crepuscular Dawn*, 153) in method, and he, like Ellul, is critical of Heidegger due to his affiliations with National Socialism. Still, as Virilio says in *Grey Ecology*, 70: "St. Augustine [along] with Heidegger spoke best about time." In Virilio's first book, *L'insécurité du territoire*, there is excellent work by Virilio in this area as he explores the relationship between *phusis* and *phuein.* Specifically, see the chapter "La politique du square," 163–178.

49. See Virilio, *The Vision Machine*, 20 for a passage that is clearly working with Augustine's famous commentary on light, darkness, the sun, and the moon in Genesis 1 and 2 as well Augustine and Thomas Aquinas's uses of "number, weight, and measure" (Wisdom 11:20) in their work. Virilio also seems to evoke a neoplatonist account of what could be described as a hierarchy of light: first, there is God, the light who creates and lights all lights; second, there is light as time, which is created by God, the light of all light; and third, the lights of the sun and the moon.

50. Paul Virilio, *Lost Dimension*, trans. by Daniel Moshenberg (Los Angeles: Semiotexte, 1984) 33 and Virilio, *Negative Horizon*, 115.

51. Virilio, *Polar Inertia*, 1; 12; and 38 and Virilio, *The University of Disaster*, 73.

52. Virilio, *Landscape of Events*, xiii. Virilio's emphasis. I am quoting Virilio's citation of the essay instead of the actual text in order to show Virilio's translation and what he wants to draw the reader's attention toward with his use of italics. For the actual text I am using, however, see Walter Benjamin, *Illuminations*, ed. Hannah Arendt, trans. Harcourt Brace Jovanovich (New York: First Mariner Books, 2019), 253–264.

53. Virilio, *Landscape of Events*, xiii.

54. Benjamin, *Illuminations*, 263–264.

55. Ibid., 253–254. Here, Benjamin describes Messianic time and its import for history that I am summarizing as follows: "A chronicler who recites events without distinguishing between major and minor ones acts in accordance with the following truth: nothing that has ever happened should be regarded as lost for history. To be sure, only a redeemed mankind receives the fullness of its past. Which is to say, only for a redeemed mankind has its past become citable in all its moments. Each moment it has lived becomes a *citation a l'ordre du jour* [citation/summons to the agenda/ order of the day]—and that day is Judgment Day."

56. Virilio, *Politics of the Very Worst*, 65; Virilio, *The Information Bomb*, 107; and, Virilio, *Crepuscular Dawn*, 156.

57. For examples see Virilio, *Art of the Motor*, 56–57; Virilio, *Open Sky*, 18; Virilio, *Politics of the Very Worst*, 87; Virilio and Armitage, *Virilio Live*, 92; Virilio, *Desert Screen*, 93; and, Virilio, *City of Panic*, 69.

58. Virilio and Lotringer, *Accident of Art*, 96.

59. Ibid.

60. Virilio and Lotringer, *Crepuscular Dawn*, 153.

61. Ibid.

62. See Virilio, *Negative Horizon*, 168; Virilio, *University of Disaster*, 54; Virilio, *The Futurism of the Instant*, 72; and, Virilio, *The Great Accelerator*, 24.

63. It is worth noting that Virilio opposes what he describes in *Ground Zero*, 10–11 as hard, limited, and exclusionary doctrines of salvation. To my mind, this means his use of general/special revelation in his work would be intended to open up rather than limit the possibly for all people to experience a special revelation of God. For example, Speaking of the effects of iconoclasm on and in the church, Virilio writes, "With Christ now occupying only a subaltern place in these new temples, believers would be returned to the ancient despair at being born human, and the act of generation would escape the *principle of indeterminacy*—the doctrine of free will

and redemption—defended, with a ferocity we know so well, by the Roman papacy." It seems Virilio defends this doctrine as well, and consequently would say that special revelation is available only to certain people and not others.

64. As far as the Old Testament is concerned, Josiah's rediscovery of the law and the reforms this inspired comes to mind (cf. 2 Chronicles 34). As for the New Testament, it says a lot to me that Jesus is constantly reminding the Pharisees and Scribes about truths revealed in the scriptures, which, for revealing reasons, they have (un)consciously or (in)conveniently forgotten. The amount of "have you not read" lines Jesus delivers in the gospels to the most religious folks of his day should alert us to the truth that the people of God are often the people most dangerously prone to forgetting the scriptures they are meant to know, love, and share with the world. For one such "have you not read" scene, see Mark 12.

65. I think this is why Virilio insists throughout his work that we should not allow technology to manipulate or replace our division, and by extension our experience, of time into the categories of past, present, and future. For example, in Virilio, *Politics of the Very Worst*, 81 he says, "The being is situated, it is *hic et nunc*. Human beings exist in the three dimensions of chronological time—past, present, and future. It is obvious that the liberation of the present—real time or world time—runs the risk of making us lose the past and the future in favor of a presentification, which amounts to an amputation of the volume of time. Time is volume; it is not only space-time in the sense of relativity. It is volume and depth of meaning, and the emergence of one world time eliminating the multiplicity of local times is a considerable loss for both geography and history." Basically, if technologies enforce a singular category and experience of time (i.e., real time) that spans the whole world, then we severely limit the opportunities to encounter any kind of revelation in our unique times and spaces (in what every situated being experiences as the past, present, or future). For other examples, see Virilio, *War and Cinema*, 60; Virilio, *Polar Inertia*, 38; Virilio, *The Information Bomb*, 118; Virilio, *University of Disaster*, 53; and, Virilio, *The Futurism of the Instant*, 71.

66. Virilio, *Aesthetics of Disappearance*, 38.

67. Virilio, *University of Disaster*, 134–135.

68. Virilio, *Pure War*, 232 and Virilio, *University of Disaster*, 124.

69. Virilio, *Administration of Fear*, 71 and Virilio, *Grey Ecology*, 43.

70. Virilio and Lotringer, *Crepuscular Dawn*, 147 and *The Accident of Art*, 88.

71. Virilio, *Politics of the Very Worst*, 89.

72. Virilio and Lotringer, *The Accident of Art*, 96. As Virilio says in respect to the duty to remember, "[Negativity] means that we remember in order not to do it again. The pyramids were preserved in order to remember the tyrannical character of the Pharaohs, who wanted to live forever. And History has been anxious to preserve this memory of Egyptian civilization. The same goes for the Greeks, the Middle Ages. As for us, in the twentieth century, we have begun to preserve the concentration camps. Hiroshima, Auschwitz are negative monuments." I think this is a unique, apocalyptic (revelatory and crisis inducing) articulation of general revelation. See also Virilio, *Grey Ecology*, 52.

73. Virilio and Brausch, *A Winter's Journey*, 72–73.

74. A comparison of Ellul, *Meaning of the City*, 176; Ellul, *False Presence of the Kingdom*, 21; Ellul, *Apocalypse*, 222; Ellul, *Ethics of Freedom*, 76–88; and, Ellul, *What I Believe*, 188–210 and 214–233 shows the difference between a pre- and post-turn Ellul.

75. See Michel Foucault, "Introduction," *The Archaeology of Knowledge and Discourse on Language*, trans. A. M. Sheridan Smith (New York: Routledge, 2002), 3–23 for a good summary of Foucault's archaeological method. As it is the case for Heidegger and Virilio, I cannot provide a comprehensive comparison of Foucault and Virilio. However, suffice it to say, and in respect to the discussion at hand, the way both thinkers use the imagery of contemplating monuments and the like is noteworthy.

76. Virilio, *Politics of the Very Worst*, 39 and 41. See also Virilio, *Grey Ecology*, 25–26 and Virilio and Richard, *Administration of Fear*, 73–74.

77. Virilio and Lotringer, *Pure War*, 143.

78. Ibid.

79. Virilio and Lotringer, *Crepuscular Dawn*, 23.

80. Emily B. Lyle, "Dumézil's Three Functions and Indo-European Cosmic Structure," *History of Religions* 22, no. 1 (1982): 25–44 outlines quite technically and thoroughly what I will only summarize in this section respect to Dumézil's tripartite social structure and its corresponding functions. Also, see Ellul, *New Demons*, 48–87 for extended discussion of the four poles of the sacred.

81. Virilio and Lotringer, *Pure War*, 22.

82. Ibid., 133.

83. Ibid., 27–28.

84. Ibid.

85. Lyle, "Dumézil's Three Functions and Indo-European Cosmic Structure," 25–44.

86. Virilio and Lotringer, *Pure War*, 27–28.

87. Ellul, *New Demonds*, 89.

88. Virilio and Lotringer, *Pure War*, 27–28.

89. I think Virilio's concerns with the overlaying and perversion of structures and functions is rooted in a Catholic understanding of creation wherein, as he says in *City of Panic*, 63, "Finiteness does not mean the end of the world, the Apocalypse. It means that an apple is only an apple, a man is only a man, and the Earth is only land." See also Virilio, *Grey Ecology*, 26. As I read him, I think Virilio is saying: if you pervert, manipulate, violate, or change the intended forms and kinds of creation—turning an apple into something other than an apple, a man into something other than a man, and the earth into someone other than the earth—then violence will result.

90. Virilio and Lotringer, *Pure War*, 28.

91. Ibid. I think there is a lot to what Virilio describes here. You can see such overlaying, or perversion, at play in the mythologized, venerated (i.e., sacred) war heroes in the United States. One such overlay of the priest, warrior, and peasant who immediately comes to mind is Chris Kyle, the Navy SEAL sniper who wrote the autobiography *American Sniper: The Autobiography of the Most Lethal Sniper in U.S. Military History*, which inspired the film of the same name directed by Clint

Eastwood. Although some of Kyle's accounts of his feats as a sniper have been disproven, he was certainly proficient at killing people in his four tours during the Iraq War. As time has passed, he has been mythologized as a titanesque combination of deadly warrior, ascetic monk, and crafty entrepreneur, who, though apparently from humble, working-class beginnings, gained a lot of notoriety and made a lot of money with his book and movie. Kyle's narrative, along with the "punisher skull" logo that he and his contingent adopted on the battlefield have, since *American Sniper*, reached mythical and sacred status in the United States. Many military and police forces, as well as far-right militias, have featured the same logo on their own uniforms, tactical gear, and vehicles. See Matthew Cole and Sheelah McNeill, "American Sniper" Chris Kyle Distorted His Military Record, Documents Show," *The Intercept*, May 25, 2016, accessed January 13, 2021, https://theintercept.com/2016/05/25/american-sniper-chris-kyle-distorted-his-military-record-documents-show/ and Sam Thielman, "How Do You Stop the Far-Right Using the Punisher Skull?" *The Guardian*, June 11, 2020, accessed January 13, 2021, https://www.theguardian.com/books/2020/jun/11/how-do-you-stop-the-far-right-using-the-punisher-skull-make-it-a-black-lives-matter-symbol for further context.

92. See, for example, Virilio, *The Art of the Motor*, 120; Virilio, *Landscape of Events*, 3; Virilio, *Negative Horizon*, 35; and, Virilio, *University of Disaster*, 126–127.

93. Virilio, *Speed and Politic*, 111–113.

94. Virilio and Lotringer, *Pure War*, 34–36.

95. Ibid.

96. Ibid.

97. Ibid.

98. Virilio, *Information Bomb*, 45–46. In fact, Virilio often uses this kind of negative Trinitarian imagery as a hermeneutic for modernity, its *myths*, and its sense of the sacred. See, for example, Virilio descriptions of the "three bombs"—atomic, information, and genetic—in Virilio and Lotringer, *Crepuscular Dawn*, 135–136 and 145–146 and Virilio and Lotringer, *Accident of Art*, 84–85; "C3I, the official designation for Control, Command, Communication, Intelligence," in *Desert Screen*, 20; *Virilio Live*, 39; and *Pure War*, 196; his descriptions of the "three types of weapons [that] have succeeded each other in history: weapons of obstruction, weapons of destruction, and weapons of communication" in *Virilio Live*, 86 and *Pure War*, 195–196; and, his descriptions of the "three phases of military knowledge," namely the tactical, strategic, and war economy phases in *Pure War*, 30. Not only do these offer powerful hermeneutics for the development of war and technology throughout history, they are also triadic concepts that form negative images of the Trinity in general, and the Triune God's divine economy in particular.

99. Ibid.

100. Which, of course, we see happening quite a lot now as wars are being declared and waged on social media platforms such as Twitter. See, for example, Riley Beggin, "Trump Threatens New Attacks on Iran, as Iranian Leaders Compare Him to ISIS," *Vox*, January 5, 2020, accessed January 16, 2021, https://www.vox.com/world/2020/1/5/21050347/trump-new-attacks-iran-soleimani-52-military-cultural.

101. Virilio, *Desert Screen*, 33.
102. Virilio, *Ground Zero*, 19–20.
103. Virilio and Lotringer, *Pure War*, 18.
104. Paul Virilio, *Bunker Archaeology*, trans. George Collins (New York: Princeton Architectural Press, 1994), 46.
105. Ibid.
106. Virilio and Lotringer, *Pure War*, 37–38.
107. Virilio and Lotringer, *Crepuscular Dawn*, 23.
108. Benjamin, *Illuminations*, 263–264.
109. Benjamin, *Illuminations*, 263–264.
110. Paul Virilio, *War and Cinema*, trans. Patrick Camiller (London: Verso, 1984), 67–68.

Chapter 4

Technology's Origins

This chapter builds on the previous chapter by moving backward in time with Ellul and Virilio, to what they view as critical moments in technology's evolution. These moments are the Industrial Revolution, Greco-Roman antiquity, the Cain and Abel narrative in Genesis 4, and the Fall account in Genesis 3. *Myth*, as it was outlined in the previous chapter, reveals the relationships between these important moments in technology's development. It creates a clash between history and revelation that forces the reader to consider their complacency in the face of, and possible complicity in, the modern world's construction, sacralization, and justification of technological idols.

THE CRADLE OF ALL TECHNICAL ACTION

Technological Society and *Meaning of the City* establish the trajectory for Ellul's fifty years of writing on technologies. Comparing key passages in these texts reveals Cain's killing of Abel and building of the first city is Ellul's theological return to zero for classic *tékhnē*, and correlatively, reveals that the Industrial Revolution is Ellul's sociological return to zero for modern *technique*. These return to zero form a dialectical relationship wherein the theological *myth* of Cain and Abel demythologizes the sociological *myth* of the Industrial Revolution, and in the process, aims to desacralize modernity's sacralization—idolatry of—technology.

Ellul begins *Technological Society* with a provisional definition of *technique* and a study of its historical development. He opens with the statement "my book is not a [full] history" of technology and "I shall speak in a historical vein [only] when it is necessary to understanding the technical problem in society today."[1] The reason for this critical approach is that "technical activity is the most primitive activity of man" and "a phenomenon which admits of no complete explanation."[2] Moreover, "Each primitive operation of man implies the bridging of such an enormous gulf between instinct and the technical act that a mystic aura hovers about all subsequent development."[3] Nevertheless, "Our modern worship of technique derives from man's ancestral worship of the mysterious and marvelous character of his own handiwork."[4] Dramatic remarks about incomplete explanations, mysticism, marvel, and worship hint at a possible theological or philosophical bridging of the gulf Ellul describes here, but in this instance, he still favors a sociological reading.[5]

When Ellul proceeds with a social-scientific reading of history, it is important that he mentions *homo faber*, as it represents both a classic and a modern concept. The concept of *homo faber* spans from Appius Claudius Caescus, a Roman politician who lived from 340 BC to 273 BC, to twentieth-century thinkers like Hannah Arendt, Max Scheler, Henri Bergson, and Max Frisch.[6] As such, *homo faber* gives Ellul a lot of historical territory to work with and indicates numerous possible locations for *myth*. So, Ellul writes, "It has not been sufficiently emphasized that technique has evolved along two distinct paths. There is the concrete technique of *homo faber*—man the maker—to which we are accustomed" and "there is [the] technique, of a more or less spiritual order, which we call magic."[7] He describes the *technical* character of magic but quickly transitions to an extensive discussion of concrete *technique* represented by *homo faber*.[8] Two important details immediately stand out in this passage. First, Ellul's mention of magic and concrete *technique* signals an engagement with Marx, particularly the Marx of *Manifesto of the Communist Party*, where he writes:

> The means of production and of exchange, on whose foundation the bourgeoisie built itself up, were generated in feudal society. At a certain stage in the development of these means of production and of exchange, the conditions under which feudal society produced and exchanged, the feudal organization of agriculture and manufacturing industry, in one word, the feudal relations of property became no longer compatible with the already developed productive forces; they became so many fetters. They had to be burst asunder; they were burst asunder. Into their place stepped free competition, accompanied by a social and political constitution adapted in it, and the economic and political sway of the bourgeois class. A similar movement is going on before our own eyes. Modern bourgeois society, with its relations of production, of exchange and of property, a society that has conjured up such gigantic means of production and of exchange, is like the sorcerer who is no longer able to control the powers of the nether world whom he has called up by his spells.[9]

Compare this to the opening chapter of *Technological Society*. Discussing the two streams of magical and concrete technique, Ellul writes, "Explanation becomes even more difficult when we note that in the magical domain too our own era has achieved an overwhelming superiority; our magical techniques have become really effective. These techniques obviously must not be confused with religious life or anything of that kind. This is purely a social phenomenon both in aim and form."[10] Like Marx, Ellul uses "magic" metaphorically to describe the character and consequences of *technique* in the modern world, just as Marx used magic metaphorically to describe *capital*. For Marx and Ellul, the putative magic of industrial machines eliminates the magic of preindustrial means. It is a hint at the sacred nature of *technique* in the modern world, but at the same time, the metaphor of magic only goes so far for Ellul. He makes a distinction in *Technological Society* between these two "streams" of early *technique*, emphasizing that concrete *technique*, however magical it appears, is only made magic through *technological* discourse. Consequently, concrete *technique* eliminated this other, more mysterious, magical *technique*.[11] Jumping ahead to Ellul's theological work, these textual connections are also signaled when Ellul similarly reworks Marx in a theological key, in the introduction to *Meaning of the City*. Here, Ellul writes:

> Cain is not the city and Abel is not the country; but the relationship between them also illuminates (not exclusively) the relationship between the city and the country. All of man's history is not limited to the history of the city and its progress. But they have nevertheless intermingled, and neither can be understood alone. The two realities are realities for God, and only in him can we know exactly what they are. But the problem becomes serious when the city kills the country, when Cain kills Abel. When that happens, man and history are so thrown out of kilter that nothing can modify the new situation. But—and here is what is important—it can be no other way. Cain could not stop being himself. From the beginning he had to kill Abel. The city, so mediocre, so puerile with its poorly carved blocks artlessly stacked upon one another, with its scanty population still rustic in nature—this city was, from the day of its creation, incapable, because of the motives behind its construction, of any other destiny than that of killing the country, where God put man to enable him to live his life as best he could.[12]

Ellul is implying Cain represents bourgeois-industrial society, whereas Abel represents feudal agrarian society. Cain's murder that "he had to" commit and "can be no other way" represents what Marx argues "had to be burst asunder; they were burst asunder."[13] All of these textual connections are probable in that this certainly is not the only circumstance where Ellul reworks Marx to make a sociological, historical, or theological point.[14] But what do all of these textual connections mean for the initial quotation from *Technological Society*? To begin with, Ellul precludes a mystified—philosophical or theological—reading of *technique's* history and privileges a demystified—historically and sociologically anchored—account of *technique's* origin and development. He does, however, leave some room for a mystified, theological, reading. The demystified account that appears in *Technological Society* begins as follows, and it hints at where another mystified reading might be located:

> Technique is essentially Oriental: it was principally in the Near East that technique first developed, and it had very little in the way of scientific foundation. It was entirely directed toward practical application and was not concerned with general theories, which alone can give rise to scientific movement. This predominance of technique in the East points up an error which is found throughout Western thought: that the Oriental mind is turned toward the mystical and has no interest in concrete action, whereas the Western mind is oriented toward "know-how" and action, and hence toward technique. In fact, the East was the cradle of all action, of all past and primitive technique in the present sense of the word, and later of spiritual and magical technique as well.[15]

These comments add historical detail to the first section of *Technological Society*'s opening chapter, where Ellul argues that *technique* originally appeared from the East, to establish a difference between classic Eastern *technique* and classic Western *science*.[16] But here, in the second section of the first chapter, Ellul elaborates on this distinction to locate geographically *technique*'s ground zero, describing the transfer of *technical* knowledge from East to West as follows: when "Greece applied herself to technical inquiry and the development of industry, she looked to the East for methods."[17] In "the first century, when Rome—the perfect example of the technical spirit in antiquity—took up industry, she too turned to the East for industrial techniques—the refining of silver and gold, glassmaking, the tempering of weapons, pottery, ship construction, and so on. All these techniques came to Rome from the East."[18] At this point, one might object that Ellul is developing a Eurocentric vision of *technique*, and it could be interpreted as such, were it not for Ellul's emphasis on the dialectical meeting of Eastern *technique* and Western *science* in a particular time and place in history that causes the mutation that eventually produces modern *technique*. As such, the modern phenomenon of *technique*, according to Ellul's historical descriptions, cannot be exclusively attributed to either the East or the West, but should be attributed to both, because classic Eastern *technique* needed classic Western *science*, and vice versa, in order for modern *technique* to develop.[19]

Ellul continues, sketching the development of *technical* activity from the first to the twentieth century, and while the details of this development are of interest, it is more important to show how the passages referenced here connect to *Meaning of the City*. Ellul begins this theological text with a description of Cain's actions after he murders his brother Abel and is marked by God: "Cain went away from God's presence and lived in the land of Nod, East of Eden."[20] This is important because, "The seed of all man's questings is to be found in Cain's life in the land of wandering, always searching for a place where his need for security might be satisfied."[21] This land is, "East of Eden. The East. The country where the sun rises, the point of departure. But one does not stay in a starting place, one leaves it. And now Cain is forever fixed at this starting place."[22] Ellul describes the significance of this geographical location as follows: "The East has an exact meaning in the scriptures. It is both the road man takes in his futile search for eternity, and the one he takes when he obeys God's call. These two great ways are parallel and show the relationship between the different attitudes of the human race."[23] A few important comments must be made in relationship to these passages.

First, Ellul's comments reflect an engagement with Augustine's two cities theology and its relationship to Cain and Abel, particularly as it appears in Book XV, especially in chapters 8, 9, 15, and 17–19 of *City of God*.[24] This also reflects engagement with Luther's writing on Cain in the first and second volumes of *Commentary on Genesis*, particularly chapters IV–VII. For instance, Ellul's mention of "two great ways" echoes Augustine's commentary, "Abel followed Cain, and he was killed by his elder brother, thus being the first to display a kind of foreshadowing of the pilgrim City of God; showing that it was to suffer unjust persecution at the hands of wicked and, in a sense, earth-born men, that is, men who delight in their early origin and rejoice in the earthly felicity of the earthly city."[25] Likewise, when Ellul writes "the seed of all man's questings is to be found in Cain's heart," he appears to be drawing from Luther's repeated emphases that Cain is "the seed" of Adam and Eve. As Luther writes, "Adam and Eve placed all their hope in their first-born only. They called him 'their treasure,' as [Cain's] name indicates. But Abel they looked upon as nothing and considered that he could do nothing; while they adorned Cain as a king and held him to be the 'blessed Seed.' From him therefore they promised to themselves great things, and of him they speak great things; and he on his part became filled with pride."[26] At the same time, Ellul differs from Augustine and Luther because he describes Cain's actions in specifically modern terms—as is especially clear when he mentions *homo faber*. Essentially, Ellul transforms traditional two cities, two realms, or two kingdoms doctrine, but he does not break away from it:

> Just as history begins with the murder of Abel—since before death there is no way for us to learn man's history, and the death that resulted in the fall first manifested itself as murder—so civilization begins with the city and all that represents. With Cain's beginning, with Enoch [the son and city], we have a sure starting place for all civilization. Paradise becomes a legend and creation a myth. Now we have something to which we can fasten our history, and the ramparts of the Canaanite or Peruvian cities give us a sure material knowledge of *homo faber*. Thus we are satisfied, and Cain was truly correct when he said that he was beginning. He is the murderer, and the one who realizes man's wisdom, who will transform *homo proto sapiens* into *homo faber*.[27]

He concludes "[Cain's narrative] fits in well with the exact meaning of 'Enoch,' which is 'initiation and the beginning of utilization.' Cain takes possession of the world and uses it as he wishes, with the goal in mind that we have already indicated. Cain creates the art of craftsmanship."[28] Here Ellul marks the theological return to zero for *technical* action. Ellul reads Cain's building of the first city as the theological representation of the origins of classic *tékhnē*. Correlatively, Ellul reads the Industrial Revolution as a sociological and historical representation of the origins of modern *technique*. The one *myth*—Cain's—is used to demythologize the other—the Industrial Revolution.[29] Moreover, modeling himself after his mentor Qoholet in Ecclesiastes, Ellul, by calling Cain the first craftsman, who "creates the art [*tékhnē*] of craftsmanship [again, *tékhnē*]," takes a Hebraic shot at Greek philosophy's (and with it, scholasticism's) *tékhnē*.[30] Which is to say, Ellul uses the scriptures to show how and why the first act of the craftsman was perpetrated by the bloody hands of a murderer instead of the dexterous hands of an artist. Furthermore, he uses Cain and Abel to theologically affirm Marx's, and by extension his own, critique of Feuerbach. As Ellul puts it, "The basis of a scientific social system (socialism in this case) cannot rest on something as vague as the feeling of fraternity. Socialism cannot take root in Feuerbach: it must be scientific."[31] This is because, for Ellul, Cain and Abel reveal how often fraternity becomes a fratricide.[32] But it is important to specify that when Ellul says Cain "is the murderer, and the one who realizes man's wisdom, who will transform *homo proto sapiens* into *homo faber*," he is not implying that Cain is responsible for the postlapsarian fracture in "man's wisdom."[33] Cain materially realizes this fractured wisdom with his building—*tékhnē*—of the first city, but the acts—*poiēsis*—of Cain's parents at the Tree of Knowledge of Good and Evil are responsible for the fracture in wisdom—*epistēmē*. As such, Adam and Eve are the first fallen *homo proto sapiens*, whereas Cain is the first fallen *homo faber* who murders his brother.

Altogether, there is an essential link with these two mythical returns to zero. Ellul's repeated emphasis in both texts that *technique*, from a sociological, historical, and theological perspective, came from the East is the link. For Ellul, the East is "the cradle of all action, of all past and primitive technique in the present sense of the word" from every disciplinary perspective.[34] As such, he only differentiates *technique*'s origins in two registers because, as a sociologist and a historian, he cannot conclude that Cain's classic *tékhnē* is qualitatively similar to modern *technique.*[35] He knows too much as a sociologist and a historian to argue in good faith that there is a qualitative equivalence between classic *tékhnē* and modern *technique*. Yet, as a theologian, Ellul can and does argue in good faith that the representation of Cain as the first craftsman (*tékhnē*) in Genesis 4 reveals an explosive theological truth about modern *technique.* Consequently, Ellul uses the theological truth of Genesis 4 to demythologize the sociological *myth* of the Industrial Revolution and desacralize its institution of the sacred order of modern *technique*.[36] True to his word, then, Ellul combats *myth* with *myth* to wage war on the idol of technology in the modern world.[37]

As far as *Technological Society* goes, Ellul's critical work with *myth* in *Meaning of the City* establishes Cain as Ellul's bridge for the "enormous gulf between [human] instinct and the technical act" alluded to at the beginning of his analysis of modern *technique*.[38] Indeed, for Ellul, Cain embodies the truth that, "Modern worship of technique derives from man's ancestral worship of the mysterious and marvelous character of his own handiwork."[39] With the myth of Cain's worship of his own handiwork—classic *tékhnē* handled apart from God—being brought into confrontation with modernity, then, another demythologizing and desacralizing blow is dealt to the modern sociological *myth* of science, history, and the sacred order of modern *technique*. Cain's theological witness dissolves the "mystic aura," which hovers over the primal instinct that sparked the first *technical* act and powers the same primal instincts that produce modern *technique* today.[40] And, reading Ellul from this perspective makes it more precise to suggest that Cain *murders* this mystic aura. Cain is the theological representation of "concrete technique" and *homo faber*. Abel is the theological representation of the magical technique "of a more or less spiritual order" briefly described and immediately stopped short by Ellul in *Technological Society.*[41] Such a two-streams/two-cities correlative reading of *Technological Society* and *Meaning of the City* explains why Ellul quickly kills off his discussion of magical *technique* in the former text. As mentioned, he demystifies rather than remystifies modern *technique*'s mysterious origins in his sociological work and to that end, he concludes, "The two great streams which we have traced from their beginnings evolved in completely different ways. In manual technique we observe an increase and later a multiplication of discoveries, each based on the other. In magic we see only endless new beginnings, as the fortunes of history and its own inefficiency call its procedures into question."[42] The significance of

Abel's name for Ellul especially exposes these respective two-streams/two-cities attributions.

In *Reason for Being*, Abel's name receives extensive commentary.[43] Ellul writes, "Abel was mist, breath, or smoke that melts away; his name predicted all the rest of his tragic life."[44] He also examines the Hebrew root of the name. "*Hebel* suggests the idea of nothingness: from the point of view of reality, unsubstantiality; from the point of view of the truth, a lie; from the point of view of efficiency, uselessness; and from the point of view of security, deceit."[45] Reading this into Ellul's comments on the "two great streams" in *Technological Society* thereby communicates how Cain's efficient manual *tékhnē* increases and multiplies, whereas Abel's inefficient magical *technique* decreases (it is murdered!) due to the tragic "fortunes of history" and an "inefficiency," which "calls [magical *technique's*] procedures into question."[46] Abel, the "smoke that melts away," is a theological way of saying what Marx said about all solid things in the nineteenth century. The name Abel, as Ellul reads it, connects to Marx's now-famous statement: "The bourgeoisie has stripped of its halo every occupation hitherto honored and looked up to with reverent awe. It has converted the physician, the lawyer, the priest, the poet, the man of science, into its paid wage laborers."[47] Now, "Constant revolutionizing of production, uninterrupted disturbance of all social conditions, everlasting uncertainty and agitation distinguish the bourgeois epoch from all earlier ones."[48] This means that "all fixed, fast-frozen relations, with their train of ancient and venerable prejudices and opinions, are swept away, all new-formed ones become antiquated before they can ossify. All that is solid melts into air, all that is holy is profaned, and man is at last compelled to face with sober senses his real conditions of life, and his relations with his kind."[49] When Ellul writes, "Abel was mist, breath, or smoke that melts away," a Marxist reading would imply Cain is the profane bourgeois-industrialist and Abel is the profaned antiquated-shepherd. Hence, Abel melts into air because of Cain, and this reading is all intricately woven together in the introductions to *Technological Society* and *Meaning of the City*.

If it appears tenuous to read Abel as a representative of magical *technique*, note Ellul's description three pages prior to his concluding remarks on the end of magical *technique*'s stream in *Technology Society*. There, he writes, "In the spiritual realm, magic displays all the characteristics of a technique. It is a mediator between man and 'the higher powers,' just as other techniques mediate between man and matter."[50] Then, one page prior, "Magical techniques, rites, formulas, and sacrificial practices disappear irremediably. The new civilization will fashion its own new stock of magic, which has little in common with the old."[51] But with manual *technique*, "When a civilization dies, it transmits to its heirs its material but not its spiritual apparatus. Tools, houses, and methods of manufacture live on and, more or less reincarnated, are to be met with again."[52] Accordingly, "Spiritual" Abel is murdered because his "sacrificial practices" are more pleasing to God than Cain's. But Cain's "methods of manufacture" live on "materially" after the first fratricide. Ellul even grants, of course, from a modern perspective "it may seem questionable; nevertheless, magic *is* a technique in the strictest sense of the word."[53] As such, "It has not been sufficiently emphasized that technique has evolved along two different paths."[54] For Ellul, those two different paths are Cain and Abel. The city of man and the city of God.

In the modern world, Abel may become less representative of the mysterious, mediatory, sacrifices of a bygone era, and more representative of a fleeting disappearance whose memory only belongs to God.[55] But Ellul reminds readers: "All is Abel; that is, condemned [to death] beforehand, just like Abel. Everything bears Abel's name. Here we see an aspect of [Ecclesiastes'] intransigence: everything that we see as power, grandeur, success—all this belongs in advance to the category of vanity. It is all condemned to disappear, to vanish, without any kind of posterity."[56] Before Marx wrote it, Abel's murder reveals all solid things and all solid people melt into air. Even the good things and the good people.

Ellul's theological account of the origins of classic *tékhnē* thereby delivers a theological hit on modern *technique*. Ellul's comments on Cain and Abel stitch together the two technical streams of *Technological Society* and the two cities of *Meaning of the City*. It explains why Ellul writes, as quoted initially, "We cannot conceive of Cain without Abel, or of Abel without Cain. They are tied in with each other, and they are tied in with humanity."[57] So, "Cain is not the city and Abel is not the country; but the relationship between them also illuminates (but not exclusively) the relationship between the city and the county."[58] In fact, "The two realities are realities for God, and only in him can we know exactly what they are."[59] With all of these connections in view, a fuller frame is formed for the first two of Ellul's four returns to zero in his dialectical writings on technologies. When Ellul asks at the opening of *Technological Society* "there is the technique of hunting, of fishing, of food gathering; and later weapons, clothing, and building. And here we face a mystery. What is the origin of this activity?" he answers his own question with *Meaning of the City*.[60] The origin is Cain, the first fallen agriculturalist and builder.[61] The origin is Nimrod, the first fallen hunter and warrior.[62] It begins after Abel is murdered and the first city of man is built by Cain.[63]

With all of the aforementioned textual connections in view, it is clear that, for Ellul, the East assuredly is the cradle of all *technical* action, which synthesizes with Western *science* to mutate into modern *technique*. Whether Ellul writes about technologies in sociological or theological terms, his reading of scripture in *Meaning of the City* and his dual use of *myth* reveals that, irrespective of the time and place, the heart of *technique*—described here as a concretization of the human heart's plans—remains consistent.[64] But, Ellul also clarifies the matter further, asserting that the external representation of this technical heart can and does change according to the time and place. He theologically exposes exactly how these changes happen. First, it is Cain in the land of Nod as described earlier. Then it is Cain's son Nimrod in "Shinar, the land of sin. No other interpretation is possible. And, after Cain, it is *the cradle of urban civilization*, Nimrod's home, the beginning of his domination."[65] In other words, the theological cradle of all technical action.[66] From this cradle of technical action, technology grows through the line of Cain. The cities Nimrod, Cain's son, builds are: "Erech, Accad, Calneh, Rehoboth, and Calah. Symbolically they are the city of eternity, the city of wilderness, the city of force. All of nature's dimensions are found there: time, space, energy."[67] These are clear allusions to *myth* and how the sacred institutes order in a society's experience of time, space, and energy. As Ellul writes in *New Demons*, "The sacred is an organization of action in a space, and at the same time it is the establishment of a geography of that space in which the action can be undertaken. It is a veritable general topography of the world, involving all aspects of the latter, material and spiritual, transcendent and close at hand."[68] Then, special attention is given to one of Nimrod's cities. "Resen, the great city, meaning 'bridle,' 'bit'[. . .]. We must not forget that the horse was the first force of nature domesticated by man and put in his service. Remember that the cavalry was one of the most fearsome of weapons. Remember for the Hebrews the horse represented power and was a symbol of an advanced civilization."[69] So, "The name 'Resen' takes on meaning as an extraordinary reminder of 'the noblest conquest made by man,' of that fantastic history of man which begins with the horse and leads to splitting the atom. For from the use of the horse for utilitarian purposes, to the use of electricity, there is a difference of science and quantity, but not of spirit and will."[70] One can discursively justify the beauty of the horse, as well as the brilliance it takes to split an atom, but such justification only ignores the fallen spirit and the will, which can make the horse brutal and the atom explosive.

Answers to questions about the qualitative and the quantitative character of modern *technique*, and how it differs from classic *tékhnē*, appear here. Quantitatively, there is a difference in the science involved in and destruction caused by the bridling of the horse and the splitting of the atom. Eventually, this quantitative difference stacks up so much that it causes the qualitative shift to modern *technique*. However, also on the subject of the qualitative, Ellul points out in the quote about the horse's bridle and the atom bomb that there is an absolutely qualitative continuity in the spirit and the will, which causes humanity to bridle the horse and split the atom. It is only the outward manifestation and the devastation caused by that spirit and will, which is qualitatively different. Ellul makes exactly this point in *Technological*

Society, writing, "It is not, then, the intrinsic characteristics of techniques which reveal whether there [has] been real changes, but the characteristics of the relation between the technical phenomenon and society."[71] Then, he gives an example. "If forty-nine shells go off in some remote place and the fiftieth goes off in the midst of a platoon of soldiers, it cannot be maintained the results are identical. To assess this change, it is not the intrinsic character of the explosion which must be examined, but rather its relation to the environment."[72] Outwardly, there is a qualitative difference in the destruction caused by the one shell that hit the platoon out of the fifty that were fired. Inwardly, there is a qualitative continuity in the human spirit and the will, which caused all fifty shells to be made and fired off in the first place.[73]

For precisely these reasons, Ellul's theological commentary on classic *tékhnē* in *Meaning of the City* is woven into his sociological commentary on modern *technique* in *Technological Society*. Ellul begins with the bridle and bit in Genesis and traces its development, and mutation, all the way up to the twentieth century's atomic bomb as described in *Technology Society*. Read in this way, Ellul's sociological analysis—which begins with *Technological Society*—shows how the qualitative material of classic *tékhnē* has changed over time due to a change in the quantity of material power that modern *technique* produces. In turn, Ellul's theological analysis—which begins with *Meaning of the City*—shows how the qualitative spirit of *tékhnē* and *technique* has not changed at all. The latter text affirms this reading and exposes the heart of Ellul's desacralizing and demythologizing set up with the *Technological Society* and *Meaning of the City* dialectic:

> When we have revealed, as here, the spiritual nucleus of a problem, we admit (and this is also singularly humble) that it is in fact the nucleus of a problem, that every aspect radiates around it. If we accept this, then what we learn to know about the city by natural means, by history and sociology, and about man in the city by psychology and the novel, must be connected, coordinated, strongly knotted together, because of the spiritual nucleus. The result is that our natural sources are dependent on revelation. I know that this will be considered a betrayal to science, but it is either this method or else the sterility of death. How is it that our historians and sociologists have not been struck with the sterility of their work? Our work of tying the threads together will not be done explicitly, for we have already traced their pattern. Our entire purpose will be to come to a decision and take our biblical information to its logical conclusions.[74]

Cain is at the spiritual nucleus of modern *technique*'s problem, modernity's idolatry problem. The material power of the Industrial Revolution radiates out of and back into this problem. If the justificatory "how" of *technique* is revealed in the Industrial Revolution's mysterious and magical machines, then the demythologizing and desacralizing “why” of *tékhnē* is revealed in Canaanite manipulations of iron and fire.[75] Neither statement nor reading betrays Ellul’s sociology, history, or theology. In fact, Ellul’s analysis of the building of the Tower of Babel in *Meaning of the City* doubly reinforces this reading with its commentary on “the problem of the name,” which is to be read, perhaps unsurprisingly by this point, as a theological rendering, but not the most important theological rendering, of the problem of *technological* discourse, *propaganda*, and their relationship to the process of *technique*.[76] These connections cannot be commented upon here but, they will be explored at a later point in relation to Ellul’s theological writings on the discourse of *technology*.

In sum, for Ellul, the Industrial Revolution and what it inaugurates in history is a time and place situated representation of what Cain reveals about technologies: the first civilization, the first city, and the first *technical* act was that of a fratricidal builder. Accordingly, Ellul’s warning is as follows: if we forget this theological *myth*, and allow the sociological *myth* of industry and revolution to form us and determine our senses of the sacred, then we risk becoming idolatrous, fratricidal builders like Cain who built a city, dedicated it to himself, and hid from God. But even though Cain’s building and naming the first city *Enoch*, meaning *beginning* or *inauguration*, is a *technologically* discursive act, it is not the first *technologically* discursive act according to Ellul. Rather, Adam, Cain’s father, was the first person to name aspects of God’s creation. Consequently, for Ellul, Genesis 3 reveals that Adam and Eve are the progenitors of *technology*. The combined threats of *technique*, *technology*, and *propaganda* in his work radiate from their postlapsarian ability to determine, and therefore justify, good and evil in separation from God.[77]

ADAM’S FIRST ACT

Adam’s first action after being created by God is naming the animals (Genesis 2:19–20). Ellul reads this prelapsarian action (*poiēsis*) as the representation of God’s gift of language and naming’s power to assign a spiritual value to who or what is named. When, after the Fall, Adam names Eve and they propagate and name their sons Cain and Abel, they hand on to Cain and Abel their precarious ability to speak, name, and justify. This sequence is the context for Cain’s naming and justifying his first city *Enoch*. Accordingly, for Ellul, the Fall reveals that *technology* is the postlapsarian discourse that justifies, sacralizes, and idolizes *technique*.

Ellul repeatedly returns to the Fall to nuance and apply it to his writings over the course of his fifty years of writing, but there is one element in his reading of the Fall, which remains constant: the justifying power of speech, which makes good and evil in the world.[78] Ellul's extended commentary on language, Adam, and Eve in *Humiliation of the Word* captures his view of prelapsarian speech, which is important for understanding his view of postlapsarian speech. Ellul writes, "When Adam speaks for the first time, he does not answer God; he names animals."[79] This textual detail is important for him because, "When [Genesis 2:19-20] says that Adam names the animals, the passage underlines the gratuitousness and ease of language. Adam does not see names already prepared, registered in advance. There is no natural science of words."[80] In other words, Adam's knowledge (*epistēmē*) of creation before the Fall is immediate and therefore unencumbered by any past or present form of science (or epistemology) in the postlapsarian world. This free, unencumbered epistemology consequently enables Adam to apprehend and name creation in a completely discerning and liberated way as God watches Adam carrying out his first acts of speech in the Garden. As for speech itself, Ellul thinks God asking Adam to name the animals reveals that "human speech comes from God; but language is made up by the human race, which decides for itself—arbitrarily—the words, the rules, and the syntax."[81] Adam's, and by extension our, decisions regarding language, speech, and naming are not without consequence, though. For, from a Hebraic perspective, "A name assigns a place and a spiritual value. By naming the animals Adam shows his power over them and puts them in their place within creation."[82] So, "[Adam] establishes his own order before the Fall: a 'taxonomy' which is a free, invented expression of the supremacy God has given him."[83] But note carefully, "In this created world, in unity and unbroken communion with God, Adam does not give a name to his wife. It is only *after* the Fall, in the midst of the disorder of powers, that [Adam] names his wife *also*: 'Eve, because . . .' (Genesis 3:20). This is a reflection of the disorder of powers."[84] Critically, Ellul emphasizes that Adam's first act (*poiēsis*) is a *speech* act; naming the animals, which assigns a place and gives spiritual value to all of God's creatures. Before the Fall, this act of naming manifests an at once freeing and ordering power over the animals; after the Fall, it manifests an at once constricting and disordering power over his wife when he gives her the name Eve.[85]

But how did this problematic disordering power come about and what does it have to do with Ellul's work on technology? A significant part of the answer to this question appears in Ellul's first extended commentary on Genesis 3 in *To Will and to Do*. The opening to the first chapter says, "The serpent said to Eve, 'You will be like God, knowing good and evil.' Such is the point of departure. Before this decision of man, Adam never raised the question of good and evil. He did not know about them."[86] It is the deceptively simple point that speech, the speech of the serpent and Eve, and Adam's lack of speech during the conversation, generates the awareness that it is possible for

Adam and Eve to deny God's Word and to steal from the Tree of Knowledge of Good and Evil. It is the connection between knowing (*epistēmē*) and the expressive act of speech (*poiēsis*) before and after the Fall, which becomes fundamental for Ellul's reading of the Fall, and as a consequence, fundamental for his theology and ethics as a whole. As he writes in *To Will and to Do*, "[What is proposed] by [the serpent] is not the possession of good and evil, as though good and evil were objects, things which Eve could lay her hands on."[87] Rather, "[Adam and Eve] only receive knowledge, comparable to that of Elohim. They are going to know that there are good and evil, that there are obedience and disobedience, that there are love and hate. They are going to know, also, that there is a Yes and a No which they can say, and which someone can say concerning them."[88] Theologically speaking, whatever is said about knowledge, must be said from a Hebraic reading of knowledge's status before and after the Fall, not a classic Greek account of *epistēmē*. Ellul puts it this way:

> What does [this] mean? Precisely that Adam, like the gods, now has the power to determine what is good and evil. So far the reasoning in this discussion has been as though the knowledge in question were of the intellectual order, as Western man understands it. To know is to understand with the intelligence what is good and what is evil. In that case we could imagine that there is a good in and of itself. But in Hebrew, "knowledge" does not have that meaning. It has rather a living, active sense. Knowing is a participation of the whole being, a commitment, and involvement. It supposes a sort of creation in the very act of knowing. And the source of knowledge is a decision. When Genesis speaks to us about this knowledge, it is saying that henceforth Adam is in possession of the decision concerning good and evil. He creates, so to speak, what is good and evil by acting. Therein resides, now, his knowledge, which at the same time makes him like God.[89]

The consumption of fruit from the tree infects Adam and Eve—their minds, their hands, and their mouths. Then, fallen Adam's source of knowledge (*epistēmē*) becomes a decision concerning good and evil, and with this decision, Adam creates (*tékhnē*), so to speak, good and evil with his actions (*poiēsis*).[90] Hence, "At this point, Adam's weakness takes on a terrifying character. God's sovereign power of discernment is indeed in his hands, but Adam's hands are not God's hands. Here now in this creation, the whole of which is cast adrift with Adam, it is he who will decide the just and the unjust, yet he himself is not just. Thus the substance of all promised knowledge is reduced at the very beginning to the discovery of his nudity."[91] Postlapsarian speech consequently becomes the primal "reflection of the disorder of the powers" Ellul refers to in *Humiliation of the Word*. From here, fallen speaking and naming—justification—apart from God literally and metaphorically make (*tékhnē*) good and evil in the world.[92] And what follows is postlapsarian history: after the Fall, Adam names his wife *Eve*, they propagate and name their twin sons *Cain* and *Abel*, and they pass on to their progeny the fallen practices of speaking, naming, and justifying good and evil apart from God. Then, after Cain kills his brother Abel, he builds, names, and justifies his first city with the name *Enoch*: the beginning, the inauguration, of *tékhnē*. As far as how such readings of the Fall shapes Ellul's understanding of *technology* and *technique*, the early text *Presence in the Modern World* lays this out in explicit terms. There, Ellul writes:

> This fact that technique is self-justifying has a theological root that is worth pointing out in passing. It is [obvious/evidence]. Genesis 3:6: "The woman saw that the tree was good for food, that it was pleasant to the sight, and that it was desirable for opening the mind. . . ." Satan uses the obvious fact as a means of persuading human beings to act, without having convinced them. One does not argue with the obvious facts. One does not argue with penicillin or with an airplane that goes more than 1000 kilometres an hour. And we have such need of this deadly, obvious fact to give us confidence![93]

For Ellul, *technology*'s justification of *technique* has a theological root. The root is Genesis 3:6. *Technological* discourse frames the apparent "good" of the fruit, which is "desirable for opening the mind" (*epistēmē*); *technological* discourse takes the "obvious fact" as a means of persuading human beings to act (*poiēsis*); and this repeatedly happens, all the way up to the supersonic plane and the effective penicillin. At its root, the spiritual nucleus of the problem is located at the Tree of Knowledge of Good and Evil, with the postlapsarian ability to determine good and evil—to justify good and evil—in separation from God. *Technological* discourse, and along with it *propaganda*, consequently justify the works-based salvation offered by the rational process of modern *technique*. As such, they are the primary means by which modernity's idol of technology is constructed. So, when Ellul argues "the one best way" of *technique*, *technology*, and *propaganda* is the "formula to which [modern] technique corresponds," he is describing the dangerous sociological, and idolatrous, obverse of a theology of salvation by faith alone in Jesus Christ, who, for Ellul, is the only way, truth, and life.[94]

Returning to *Humiliation of the Word*, Ellul cites his early commentary on the Fall in *To Will and to Do*, focusing on the severance of hearing the truth of God's Word from seeing reality for what it is in light of God's truth. For Ellul, the severance between hearing truth and seeing reality eliminates the immediate relationships of God and people and inserts endless mediations into all relationships within postlapsarian creation.[95] Reiterating in *Humiliation of the Word* what he says in *To Will and to Do*, Ellul writes, "We can foresee the Fall as soon as sight is considered independently of the word; and this sight refers to the truth, appropriately enough! In Genesis 3 we read that 'the woman *saw* that the tree was good for food, and that it was a delight to the *eyes*, and that tree was to be desired to make one wise' (Genesis 3:6). This is the first time sight is a separate issue."[96] But then he moves his analysis further in *Humiliation of the Word*, saying the Fall marks the departure point for all of history, science, and *technique*—the two primary modern *myths*, which institute the sacred order of technology in the world. It is worth quoting Ellul at length here, given the importance of what he says in the passage:

> It is a considered a blessed disobedience, a blessed "Fall," that gives birth to History and Science. Finally! [. . .] The historians, exegetes, hermeneutics experts, and philosophers who talk this way never realize that they themselves could be the stupid and absurd ones! [. . .] Their eyes were opened with respect to what? Their nudity. That is the great first discovery. And that is all. All their new sight allows them to discover is this nudity. And this discovery leads them to make clothing for themselves. But this has been explained as a moral matter: they are ashamed of being this way. What we have here is moral conscience and the discovery of "modesty," or the myth of the origin of clothing. Biblically, nudity is essentially the sign of fragility and weakness [. . .]. Clothing is the sign of and search for protection. This is the essence of this marvelous discovery, thanks to their opened eyes: humanity is weak and without protection or defense. Furthermore, Adam's first reaction when he hears the voice of God (in the following verse) is fear. Thus the eyes being opened in no way constitute a point of departure for the luminous ascent of humanity. In no way does it open the possibility for humanity to understand, to assert itself, and to begin history. Instead, this event amounts to the discovery of the terrible reality that humanity without God is nothing. Without [God] we are without protection, without strength, and we are fragile as a breath. Sight reveals our effective reality to us. Thus we can certainly say that this event will be the point of departure for "science and technique," since these are essential means for escaping from this weakness and acquiring protection and power. We can certainly say that this will be the point of departure for history, since it is composed of all humanity's efforts to subsist and persist. But there is nothing grandiose and sublime in this.[97]

If searching for Ellul's "spiritual nucleus of the problem" mentioned in *Meaning of the City*, look no further than the Fall.[98] The Fall is Ellul's final return to zero and it critically frames his dialectical writings on *technology*, *technique*, and *propaganda*. Accordingly, Ellul writes in *Théologie et Technique*, "The first time the Bible speaks about technique, it is concerning Cain, although nothing prevented the Bible from speaking about it concerning

Adam as well."[99] Regardless, "By whatever means we take technique, we realize that it can only be a phenomenon of the Fall, that is nothing to do with the order of [prelapsarian] creation, that it does not result in any way from the vocation of Adam wanted by God, but necessarily from the situation of fallen Adam."[100] Ellul does not equate Adam and Eve's fashioning of clothes out of leaves after the Fall with *tékhnē* because the postlapsarian speaking and naming of Adam and Eve, and how it determines thought (*epistēmē*) and action (*poiēsis*), concerns Ellul much more than the concrete act of *tékhnē*. How else is it that Cain learned to name and justify apart from God the first city "Enoch?" How else is it that the builders of the Tower of Babel said, "Come, let us build. . ." in order to "make a name" for themselves?[101] *Technology*, the postlapsarian justifying discourse of *technique*. The East, just East of Eden, is the cradle of all *technical* action for Ellul's work technology.[102]

Although the Fall reveals that *technology*, for Ellul, is the postlapsarian discourse that justifies *technique* in a world separated from God, this does not mean Ellul condemns *all* human speech and naming after the Fall. Rather, he condemns a *certain kind* of human speech and naming after the Fall: the kind that justifies power, domination, and violence with the naming of atomic bombs and proclamations of the putatively salvific potential of consumer products.[103] For Ellul, speaking and naming—language—are powerful, beautifully powerful when used with care, but also brutally powerful when misused. This means the postlapsarian speaking and naming—the language—of fallen creatures is precarious and precious, potentially destructive and constructive. As Ellul puts it, "Truth is communicated by the most uncertain means, the one most prone to variations and doubt, as we have seen—by the word, that fragile thing that does not last, evaporating as soon as it has been said. Thus what we are surest of is connected with the most uncertain thing in existence."[104] *Did God really say?* the serpent asks. Adam and Eve allow themselves to become uncertain about that which they must have been most certain: the Word of God given as a gift in the Garden.[105] A lot can be learned, then, from the uncertainties of Adam and Eve.

With the aforementioned context in view, it becomes evident why, from early to late writings, Ellul constantly laments and warns about *technology* and *propaganda*'s justification of *technique*—along with the justification of anything or anyone apart from faith in Jesus Christ. For example, he writes in *False Presence of the Kingdom*, "This determination to justify modern man is the opposite to proclaiming his justification in Christ. [What] methods of justification introduce is the manoeuvre which man over and over again tries to use: that of declaring himself righteous independently of Christ; and here the theologians provide modern man with the means of declaring himself righteous in his works, in his technology."[106] Or, later, in *Hope in Time of Abandonment*, "The Church's great concern is to justify what man and the world are doing. This is just like olden times, except that it is no longer the same things which are being justified. Instead of the royal power, the patriarchal family, feminine servitude, world morality, etc., the things now being justified are science, technology, world management, politics, and the big city."[107] In fact, theology often "offers the public a justification for what is going on in the world, for what man is in the process of doing. It is true that modern man in his most fallen aspect wants exactly above everything else that someone should come along to tell him that he is right in doing what he is doing [and] that was [and is] the springboard for all the propaganda."[108] Finally, and succinctly, "Here again, for the hundredth time"—perhaps the thousandth—"I call attention to the fact that to condemn the act of justifying, is not, to my mind, the same as to condemn what is being done in society, in science, and in technology. I have never condemned [society, science, and technology]!"[109] It could be argued based on significant textual evidence, then, that Ellul's fifty years of writing about technology in both his sociological and theological texts represent a consistent attempt to demythologize, desacralize, and demolish a modern works-based theology of salvation manifested, mythologized, and sacralized in the modern world by the triad of *technique, technology*, and *propaganda*; and crucially, that *technology* powerfully and pervasively justifies the triad.

But is *technology*, the postlapsarian justifying discourse of *technique*, really *that* powerful; powerful enough to *make* evil in the world? Ellul thinks it is. In *To Will and to Do*, he writes, "Technology offers man a fulfilment of the good which is easy, effective, and justified in advance. Man's decision is obtained through adhering to technical progress. There can be no debate, no personal decision involved in the matter. The good is obvious. It goes along with power. There can be no question of escaping it. One will end with a perfect conformism never yet achieved in the moral sphere."[110] Or, in *Ethics of Freedom*:

> We no longer bow down physically to our works today. We do not kneel to an airplane nor address a prayer to it [. . .]. We do not offer incense to television sets and other gadgets. On the other hand we ascribe saving significance to it. As regards the first point the phenomenon is a familiar one. Work which is purely material and which has results that are purely material, although by no means negligible, since there can be no contesting the value of technical progress in its own sphere, this work is triumphantly transformed by man into a value, i.e., into a criterion of good and evil, into a significant aim in life without which life would not be worth living. The work of our hands is what justifies life and gives it meaning. Instead of being a purely useful and utilitarian affair which is modestly kept in its own place, it takes on the significance of truth or justice or freedom. It is itself truth, justice, or freedom. We assign to it that which really

belongs to God. In exchange for the materials blessings it confers on us, we attribute a spiritual value to it.[111]

The primal impulse to justify and idolize the works of our hands in and through speech throws us back to the Fall, where it is revealed that speaking and naming so often become a verbal "reflection of the disorder of the powers" that can, and often does, make evil in the world.[112] For this reason, the self-justifying *discourse* of *technology*, which started with Adam and Eve's *propagation* of Cain and Abel after the Fall, is Ellul's most important return to zero for *technology*. It is theological root for Ellul's identification, exposure, and dismantling of the dangerous idolatry of technology in the modern world. More will be on this in the next chapter. But before that, the next two sections will show how Virilio also sets up a clash between history and revelation in his work on technology and also uses Industrial Revolution, Greco-Roman antiquity, Genesis 3, and Genesis 4 in the clash.

THE CITY'S ORIGINS

Like Ellul, Virilio thinks the Industrial Revolution and Greco-Roman antiquity are events on history's landscape that generated modernity and its idolatry of technology. He also uses the Cain and Able narrative in Genesis 4 and the Fall account in Genesis 3 to set up a clash between biblical revelation and these historical events. Similar to Ellul's work on technology, this creates a dialectical confrontation in Virilio's writing, which is intended to expose and dismantle modernity's technological idolatry. But Virilio is not as exact and exhaustive as Ellul is when describing the Industrial Revolution and Greco-Roman antiquity. He tends to write about these historical moments in concise, conceptual, and mythologized terms, typically emphasizing granularities in these broad historical events in order to lay bare the bloodier, idolatrous aspects of modernity's history and the technologies it brought into the world. Virilio also does not separate into different texts his dialectical clashes between the Industrial Revolution, Greco-Roman antiquity, the Cain and Abel narrative, and the Fall account. Instead, he uses the fragment or constellation method discussed in chapter 1. With such an approach, Virilio remythologizes scripture and theology in ways similar to Ellul, but he tends to forms accounts of history that vacillate between demythologization and remythologization. He does this to force the reader to grapple with a reality and a truth that registers in the modern identity and consciousness as *myth*, and in the process, experience a demythologization of the *myth* wired into their own identity and consciousness.

For Virilio, the reality is that the Industrial Revolution, and with it modern technology, were set into motion by innovations in war technologies; the truth is that moving backward, from the Industrial Revolution, to Greco-Roman antiquity, to Genesis 4, reveals that the city, and with it civilization, was built with bloody hands, on bloody soil; and finally, moving back even further to Genesis 3, the start of the human race (*drómos*) is revealed: the primal scene of humanity's Fall. With this in view, this section examines the connections between the Industrial Revolution, Greco-Roman antiquity, and Genesis 4 in Virilio's work on technology. The next section, which is also the conclusion to this chapter, will show how Virilio, like Ellul, has the Fall account in Genesis 3 at the heart of his identification, exposure, and dismantling of modernity's idolatry of technology.

Virilio often uses the terms *Industrial Revolution*, *Industrial Transport Revolution*, and *Industrial Transportation Revolution* interchangeably. He plays with the term commonly used to describe an important moment in modernity's history to show how the Industrial Revolution, however much it may have shaped modernity, is only one of many examples of an enduring human tendency toward speed throughout history; and, how this enduring tendency drives humanity to seek technologies that will make the speed more powerful and further reaching. As he writes in *City of Panic*, "The great enigma of history, is *the transport revolution*, which survives and transcends not only the Industrial Revolution, but the information revolution, with its highways, its peripheral feeders or slip roads, [and] its proliferating of underground networks."[113] Basically, he says, if technology's development throughout history is considered with the hermeneutic of speed and transport, a consistent drive for speed and power is revealed. As Virilio points out in this quote, the human drive for speed and power not only exists and endures throughout history, it also steadily gains momentum, reaching peak momentum with modernity. So, the image of ever-developing roads and highways he uses in this quote are technological parts, which represent the whole human inclination toward speed and power as it exist, endures, and grows throughout history.

For this reason, Virilio thinks the Industrial Revolution is an acute instantiation of human drive for speed. But the Industrial Revolution is neither the origin nor the end of this tendency. Still, the Industrial Revolution is important, because it marks the moment in history when, through significant innovations in war technologies inspired by the human desire for speed and power, the military structure and warrior function start to pervade the sacred/priest and economic/peasant structures and functions. With time, this militant perversion causes modernity to sacralize and monetize its growing war machine, thereby making the peasant and the priest functions increasingly driving by the warrior function. As seen in a quote from a previous chapter, Virilio thinks "all of us are already civilian soldiers, without knowing it. And some of us know it. The great stroke of luck for the military class's terrorism is that no one recognizes it. People don't recognize the militarized part of their identity, of their consciousness."[114] The reason for this militarization of modernity's identity and consciousness is the speed and power attained during the Industrial revolution. Such speed and power has only increased since that time. This is why, as Virilio puts it in an interview, the innovations in war technologies during the Industrial Revolution helped "[install] speed as the categorical imperative of the modern world—including in everyday life."[115] So, once this categorial imperative is established in the modern world by faster, and more violent, technologies that appeared in this particular event on the landscape of history, the military structure and the warrior function eclipses the other structures and functions, and the idolized war machine becomes the engine that mobilizes a terribly violent and bloody early, middle, and late modernity. Elsewhere, in *Negative Horizon*, he describes how the installation of this categorical imperative of speed took place during the Industrial Revolution. There, he writes,

> "As the art of moving armies," logistics would therefore seem, Werner Sombart noted in 1913, to arise with industrial concentration and specialization. It is the economy of destruction, along with mass production, that demands the calibration of firearms (cannon, musket) and the standardization of the bore of pipes. Then it demands the standardization of munitions and, therefore, indirectly, ballistic performance and range with the cylinders of machine motors. However we look at it, the reference to war as the origin of the rise of technology is clear: in 1673 it is the gun, internal combustion engine of one cylinder, that will serve as the modern for Christian Huygens' gunpowder engine. The definition of which will later become known as the internal combustion engine is significant: "In my engine," writes Huygens, "The violent effect of the powder is reduced to a movement that can be controlled."[116]

Sombart was a German economist and sociologist, who, in the twentieth century, theorized late capitalism and the creative destruction associated with various modes of capitalism.[117] In this passage, Virilio uses Sombart and particular historical details embedded in the Industrial Revolution to advance his thesis that war technologies set the Industrial Revolution into motion, and with it, laid the tracks for modernity and put the sacralized war machine on its rails. There may be historians who would object to Virilio's suggestion that military rather than economic imperatives precipitated the Industrial Revolution and its technological innovations. Fair enough. But Virilio and Ellul would respond with the point that it must at least be acknowledged that war and economic imperatives were far more coupled during the Industrial Revolution than most people think they were, and for this reason, such an undeniable historical coupling can reveal important truths about modernity and its past and present technologies.[118] In fact, as a proper historian who provides more detailed accounts of the Industrial Revolution than Virilio does, Ellul takes this as such an established fact among historians that he does not dwell too much on it in his work.[119] Virilio, on the other hand, frequently emphasizes the connections between militarism, the Industrial Revolution, and the technologies it brought into the world. He does this, it seems, because he thinks that if he does not, nobody else will. Still, as suggested earlier, Virilio does not view the Industrial Revolution as the origin of the human drive for speed and its pursuit of technologies that will increase its power and reach. In fact, he is like Ellul, in that he looks back in history and sees pronounced manifestations of this human tendency in Greco-Roman antiquity, the same time and place as Ellul's ground zero for classic *tékhnē.*

Virilio differs from Ellul in method when he examines Greco-Roman antiquity, however, because he describes it in ways quite similar to his descriptions of the Industrial Revolution as shown earlier. Similar to those passages, Virilio blends transport and war imagery to describe an era that reflects significant acceleration in technological innovation that, just like the Industrial Revolution, was motivated by the inherent human drive for speed and power. First, he says in *Virilio Live* (note especially how he uses transport imagery to capture the conflation of the military structure and warrior function with the economic structure and the peasant function):

> If time is money, as they say, speed is power. You see it with the velocity of the predators, of the calvary, of railways, of ships and maritime power. But it is also possible to see it with the velocity of dispatching information. So all my work has been about attempting to trace the dromocratic [i.e., speed and race] dimension of societies from ancient Greek society right up to our present-day societies. This work is of course about unrelenting acceleration.[120]

Second, he says in *Crepuscular Dawn* (again, notice in this passage the transport imagery used to reflect the military/warrior structure and function's perversion of Dumézil's three structures and functions):

> Technology colonized external space first by means of the Roman roads, of the great canals, of high-voltage wires, infra-structures, and it is now colonizing the animal body [. . .]. The first mortality occurs when the relation between the body proper and the world is cut off [. . .]. The territorial loss, in this respect, cannot be compared to the one that results from the revolution of a social body—monarchy, republic, democracy, tyranny, etc.—because it is the loss of the body proper, the geophysical body [. . .]. Speed here is essential. The temporal compression [results] from the power of instantaneous speed, of interactivity, of feed-back, etc. [. . .]. As early as the Roman Empire, as early as the roads and the canals, and then railways and highways, the vocation of technology by virtue of its globalization

> dimensions was to irrigate the territorial body. Then they irrigated the social body, specifically with electricity and the energy revolution: whence the telephone and everything that has come after it. And the city was the place of this irrigation.[121]

Both quotes add further breadth and depth to Virilio analysis of humanity's drive for speed and power that gains velocity on history's plain. From the predator, to the calvary, to the train, to the boat, to data packets flung from router to router around the globe; from the first Roman roads and canals, to Industrial Europe's first railways and highways, to the globalized world's information networks, the transport infrastructures developed and enhanced throughout history are symbolic parts of the whole human tendency to pursue speed and power and the kinds of technologies that will increase gains in both. The darker side of this whole human tendency that Virilio exposes, of course, is the violence and war that this drive for speed and power produces. Further, it does not matter if the violence and the war is intended or not, Virilio points out. It is, rather, an unavoidable part of the whole human drive for speed and power. If the drive is allowed to move forward unchecked, so will violence, war, and technologies that make violence and war more brutal and expansive. In these two quotes, the violence and war are being enacted by a form of colonization that persistently irrigates the physical world with faster, further and deeper reaching, and consequently more powerful and violent, technological infrastructures. Likewise, the passages quoted earlier show how the explosive violence of the first rifle is mirrored by the controlled, but still explosive, violence of the first combustion engine. What this means, essentially, is that Virilio thinks Greco-Roman antiquity and the Industrial Revolution are critical factors in modernity's genesis, and they shed light on modernity's out of control drive for speed and power, its idolization of the war machine, and the violence this causes. But he does not think either is the origin of the drive and idolatry. So if Virilio does not locate the origin of this human drive in the Industrial Revolution or Greco-Roman antiquity, where does he locate it?

As suggested at the beginning of this section, Virilio locates part of this drive and idolatry's origin in Genesis 4, the Cain and Abel story. So it is here that history and revelation start to clash in Virilio's work on technology. First, the following passage from *Ground Zero* shows Virilio's fragment/constellation method in play, indicating how Virilio usually confronts the Industrial Revolution and Greco-Roman with biblical revelation, particularly the Old Testament, and more particularly Genesis and its account of humanity's seizure and surpassing of knowledge of good and evil:

> For most of its enthusiasts, trusting in progress would come down to believing in a beyond of *good* and *evil*—not so much that of a late Nietzscheanism, as the *tov* and *ra* of Judaeo-Christian Genesis and its old—more tribal than

> moral—laws. This would explain the appearance of new popular heroes capable like the scientific hero, of destroying themselves by spectacularly flouting *any prohibition*, any limit, any law. From the eighteenth century onwards, these standards would be offered to the credulity of the public by the young industrial press, and the adulation of the masses would go to the revolutionary *abnormality* of a new kind of women and men with, at the top of the hierarchy, the record breakers in every field: sportsmen, soldiers, adventurers, engineers, sailors—all those who make the front page of the tabloids by showing themselves capable of extreme performances.[122]

In this passage, there is a constellation of early, middle, and late modern historical phenomena and allusions to the Book of Genesis, specifically its revelation concerning *tov* and *ra*—or, the contrasting inclinations and drives of human creatures. On the one hand, there is *Yetzer ha Ra*, the human inclination and drive to commit evil acts through the violation of God's desire for creation and creatures. On the other hand, there is *Yetzer ha Tov*, humanity's potential to commit good acts by living in conformity with God's desire for creation and creatures. By using these theological motifs, a biblical reference and a theological concept is added into Virilio's constellation of modernity's history and its manifestations of the human drive for speed and power. Altogether, the constellation reveals that the human inclination toward—or worship and idolatry of—speed, power, and the technologies this inclination (worship) produces is connected to the competing drives of *tov* and *ra* in God's creatures. As a result, the constant human drive toward speed and power is revealed by Virilio's use of scripture and theology to be produced by an even deeper, post-lapsarian drive to violate God's desires for creation and creatures.

In *Speed and Politics*, these deep human drives are revealed even further when Virilio uses the Cain and Abel narrative in the chapter "The Boarding of Metabolic Vehicles." There, Virilio writes the following: "For the Judeo-Christians, everything is said from the very first pages of the *Bible: the warrior is a perverted priest*" and the "deterritorialized farmer, the first killer of men is immediately designated a *builder of cities* (a commoner)."[123] For Virilio, Cain, the first farmer, murderer, and builder of cities, whose attempt at a priestly sacrifice was rejected by God when compared to Abel's priestly sacrifice, is Virilio's revelatory *mythical* origin of the military structure and warrior function's perversion of the sacred/priest and economic/peasant structure and function. Similar to Ellul, Virilio uses the Cain and Able narrative to reveal the origins of the deeper tendencies within humanity that drive it away from *tov* and toward *ra*. In this case, the deeper human drive toward *ra* produces a strong tendency toward the speed and power of modernity, its technologies, and its worship of the war machine.

With the aforementioned context in view, the theological motifs in Genesis 4 that Virilio remythologizes in *Speed and Politics* amplify aspects of Ellul's work in *Meaning of the City*. For Ellul, the bloody hands of the marked man who built and named the first city reveals significant theological truths about classic *tékhnē* and modern *technique*. For Virilio, the bloody soil crying out to God, and Cain being driven away from the ground, reveals significant truths about the created territory upon which the first city is built. There is a seven-year gap between the publications of *Meaning of the City* (1970) and *Speed and Politics* (1977). Given the similarities between both texts' remythologizations of the Cain and Abel narrative, it suggests that Virilio read and was inspired by Ellul's doctrine of the city as he was beginning work on *Speed and Politics*.[124] There are no citations in the original as well as the translated editions of *Speed and Politics* to confirm this textual connection, but given the timing of publication, and given Virilio's highly similar remythologization of Genesis 4, there is a possible influence. Whereas Ellul remythologizes Genesis 4:17–18 by focusing on Cain and how his bloodied hands built (*tékhnē*) the first city and how he named that first city *Enoch*, Virilio focuses on how Abel's blood in the soil cries out to God and how Cain is driven away from the ground in Genesis 4:11–14.

First, based on Virilio's reading of Abel's blood in the soil crying out to God, Virilio believes creation cries out to the people who God enables to perceive its speaking and crying. This reading and application of Genesis 4 forms a revealing theological frame for Virilio's understanding of the landscape of events, and the gazes of God and Benjamin and Klee's angel upon that landscape, as discussed in chapter three. As one of the passages cited in that chapter shows, Virilio writes, "Apropos the false wit of these erudite demiurges who can tell so much just by looking at the Earth and the sky, but are, on the other hand, quite incapable of telling the time we're living in, let's turn to the Psalmist again: 'Truth shall spring out of the earth; and righteousness shall look down from heaven.' "[125] With Virilio's reading of Cain and Abel in view, this passage takes on deeper, darker meaning. For Virilio, history's blood-soaked soil helps truth spring up from the earth, as God watches from above, and eventually comes down to solve the problem with an offering of *his* blood. Accordingly, the bloodied soil of history is crying out and revealing truths to God and humanity, and Virilio prompts us to ask with his use of the Cain and Abel story: is anyone listening to or looking at what is in and upon the soil? As negative as such imagery and questions can be, there is nevertheless a positive correlative truth contained in the imagery. The bloodied soil, with the cities of blood built upon it, certainly is crying, but its cries and its sediments are "shot through with chips of Messianic time."[126] Likewise, the cities may have closed their gates to God, but "every second of time" is still "the straight gate through which the Messiah might enter."[127]

Creation cries, but it cries out the truths that creatures need to hear. Creation and the creatures who live in it need a Messiah.

The theological theme of creation's speech appears elsewhere in Virilio's writing, most obviously in an early and untranslated essay from *L'insécurité du territoire* (*Territorial Insecurity*), titled "La politique du square" ("The square's politics").[128] In this essay, Virilio describes an "imperfectly called" natural morality in the physical world (*phusis*), which is unwanted by most people but belongs to a scattered and small tribe.[129] It is an allusion to Deuteronomy 4:27, which, depending on how the Hebrew word *nahog* is translated, reads: "The Lord will scatter you among the peoples; only a few of you will be left among the nations where the Lord will [*lead/drive*] you." Bringing these theological motifs into Virilio's remythologization of the Cain and Abel narrative in *Speed and Politics*, it means Abel's blood in the soil can also cry out to the people whom God enables to perceive that form of speech.[130] It also means that Cain, the one who is driven away from the ground and retreats into the confines of the city, theologically frames Virilio's *dromological* analyses of technologies, which can harness motion (*kīnēsis*) and enhance humanity's racing (*drómos*) away from God, and away from the environment of creation (*phusis*) in which *epistēmē*, *poiēsis*, and *tékhnē* all take place. For Cain, it is a race toward a city, which eventually becomes a tomb, and all throughout history, similar races are run, and similar drives for speed and power are given control.[131] With this in view, the full passage from *Speed and Politics* quoted earlier warrants a focused look:

> The conqueror, the warrior claims a function that seems to be a perversion of the priest's. For the Judaeo-Christians, everything is said from the very first pages of the *Bible: the warrior is a perverted priest.* In fact, the first murder revolves around the rent received by God in exchange. God willingly accepts the sacrifice paid by the shepherd Abel but refuses that of the farmer Cain. The image of the first killer of men bears a direct relation to *the rent paid on the earth*, and here again everything is said in several lines: the *suffering* of the soil "opening its mouth" and crying out as it drinks blood for the first time—the blood of the territorial body as it slips away. (It shall no longer yield to you its fruit, says God . . . you will be a *wanderer* crossing over and invading the earth). Deterritorialized farmer, the first killer of men is immediately designated a *builder of cities* (a commoner).[132]

Read with passages like these in mind, Virilio's remythologization of the Cain and Abel narrative can be seen as theological frame for his constant analyses of the connections between technology and war. For, Virilio repeatedly states in writing, and in interviews, that he believes the city's origins are in war.[133] Passages like these from *Speed and Politics* consequently give readers a scriptural and theological referent for his understanding of those origins. Which is to say, given his remythologization of the Cain and Abel narrative in *Speed and Politics*, it is evident that Cain, who Virilio calls in this chapter a *warrior* and a *perverted priest*, theologically characterizes his studies of wars, which are declared on people (Abel) and God's creation (blood in the soil).[134] An allusion to Psalm 118:22 and Matthew 21:42 also appears in this remythologization passage because Virilio refers to Abel's blood paying a rent, which is due for the territorial body of God's creation (Abel's blood first, Christ's blood second).[135] The implications of this allusion will be elaborated on in a later chapter, but here it is sufficient to conclude with a look at how Ellul and Virilio similarly remythologize Genesis 4 in their respective writings on technology.

When read together, the ways in which Ellul and Virilio remythologize Genesis 4 develops a thick account of ethical thought, action, and making in the world, whether thought, action, and making are referred to via their modern or classic cognates. For Ellul, Cain reveals why *tékhnē* can no longer be thought of or spoken about in the classic sense, and further, why the sociological *myth*, which institutes the sacred order of technologies in the modern world ought to be demythologized and desacralized. For Virilio, Cain reveals why people ought to pause long enough to sense the environment (*phusis*) in which thought (*epistēmē*), action (*poiēsis*), and making (*tékhnē*) take place, because that created environment is crying out and groaning to God and God's people.[136] Also, just as Ellul thinks Cain sits at the spiritual nucleus of the problem of modern *technique*, Virilio thinks Cain is a biblical figure who reveals the deep drive within humans that pushes them away from *tov*, toward *ra*, and consequently inclines us to pursue and worship speed, power, and the technologies, which accelerate both. At the same time, Virilio's remythologization of Genesis 4 is not the only part of the constellation he forms with scripture, theology, Greco-Roman antiquity, and the Industrial Revolution. Nor is it the only theological frame for his combat with modernity's idolatry of technology and its sacralization of the war machine. Like Ellul, the Fall account in Genesis 3 is a critical part of Virilio's constellation of and frame for modernity's history and its idolatry of technology.

THE START OF THE HUMAN RACE

Virilio's reading of the Fall strengthens Ellul's reading of the Fall as described in this chapter. The fall kicks off the human race (*drómos*) when Adam and Eve, along with the rest of the human race, begin to move (*kīnēsis*) faster and further away from God and God's creation (*phusis*), progressively disrupting all created bodies and senses, from the start of the human race until now. Lucifer, the manipulative bearer of bad light, causes Adam and Eve to look at the fruit of the Tree of Knowledge of Good and Evil in a different light than the light of God. Consequently, Virilio is wary of the putative light (*phōs*) of reason.

Ellul is upfront about his reading of the relationships between sight, the Fall, and sin, especially in *Humiliation of the Word.* For instance, he observes, "Sight played an enormous role in the Fall and caused all of humanity and language to swing to its side."[137] So, "It is understandable that the Bible so often relates sight to sin [because] the eye becomes the link between reality and the flesh. The eye is seen as the focusing lens of the body (but only of the body). The Bible speaks of the lust of the eye and the eye as the source and means of coveting."[138] Such readings are central to Ellul's separation of sight/reality and hearing/truth in *Humiliation of the Word*, as well as other texts, but his allusion to Matthew 6:22–23 and Luke 11:34–36 especially is important to note. It marks a primary difference between how Ellul and Virilio approach sight, light, and the body. Ellul concludes, quite dismissively in this quote, that the eye is "only" the focusing lens of the body. In contrast, Virilio as a consequence of his reading of the Fall, frequently focuses on the relationship between sight, light, and the body.[139]

Virilio agrees with Ellul but only to a certain extent. Sight's corruption does play a significant role at the Fall and many scriptures warn that seductive sights often lead a person to sin.[140] But Virilio proceeds from his agreement with the scriptures and Ellul to focus intently on sight's role at the Fall to show how the serpent's seduction of Eve and Adam sparks "a phenomenon of consciousness," which causes "a brutal disruption of vision" and completely changes the appearance of the world. Accordingly, he moves his analyses of sight, light, and the body in a different but complementary theological, ethical, and philosophical direction.[141] For Virilio, the serpent's seduction—he uses the Latin term *seducere*, meaning, *a leading astray*—marks "a major departure for humanity, the beginning of a navigation of body and sense from something immovable toward another category of Time, a space-time essentially different because it is sensed as [unstable], mobile, conductive, transformable, like the creation of a second universe."[142] In other words, if Virilio has a theological return to zero for his understanding of the departure point for science and history, and if Virilio has a theological return to zero for his understanding of the human drive for technologies that can increase speed (*dromós*), harness motion (*kīnēsis*), manipulate light (*phōs*) and alter senses and the environment (*phusis*) in which *epistēmē*, *poiēsis*, and *tékhnē* take place, it is, just like Ellul's, the Fall.

Virilio does not read the Fall as the origin of humanity, however, because this comes before the Fall in Genesis 1 and 2. Instead, he reads the Fall in Genesis 3 as the origin of the human *race* (*drómos*).[143] As the previous quote demonstrated, Virilio sees creation as "immovable" before the Fall to the degree that Adam and Eve were created and placed within stable creation, were close to God, and their minds, bodies, and senses were aligned with God and creation. But, after the Fall, the human race (*drómos*) kicks off and Adam and Eve, along with the rest of the human race, begin to move (*kīnēsis*) faster and further away from God and God's creation (*phusis*). Then, all created bodies and senses, from the beginning of the human race's history until now, increasingly are mobilized, disrupted, fragmented, and disappeared. So, when Cain murders his brother, bloodies creation's soil, and runs toward the confines of a city after being driven from the ground and away from God, he is a noteworthy participant in the human race (described in the previous section). But as for the human race's start, it was Cain's parents, Adam and Eve, who kicked it off at the Fall. From this biblical and theological perspective, it is revealed, then, that not all idols are immobile. In fact, as Virilio shows how, according to Genesis 3 and 4, many of the most powerful and worshipped idols are incredibly mobile. Especially in modernity. Virilio writes in *Negative Horizon*, "The technological motor resulted in the long-standing pursuit of the *perpetuum mobile* [perpetual motion], and with it the release [of] violence. Two questions present themselves: How did we ever guess at the vehicle within the animal? The motor in their limbs? How did the primate come to have this desire to couple with the mount? What sort of seduction is at work here?"[144] For Virilio, Genesis 3 reveals the source of the seduction: the start of the human race and its drive toward the seduction of *ra* and away from the wisdom of *tov*. We can see this especially at play in a passage like the following (it is worth quoting at length given its importance for understanding Virilio's work on technology):

> This beloved, who, according to Novalis, is *the microcosm* of a universe which is only *an extension* of the loved one, the body of the woman becoming one with a communication body, is the ideal vector between man and the new world—it's no longer simply a matter of a couple, but a sort of triangle. The solitary movement of *seducere* or sexual coupling necessitating solidarity in movement—the couple is also a yoking, a coupling, the constitution of an effect of common discipline, a kind of two-seater vehicle, implying the territorial body as a third partner. In the episode of Genesis we observe that the rites-of-passage from one world to the other provoke not only a metamorphosis of vision but also an immediate dissimulation, a prudent camouflage of bodies. We may recall here the reflection of Hannah Arendt: terror is the end of the law of movement. In [Genesis 3] fear is contemporaneous with seduction because [seduction] is precisely production of the exceptional phenomenon of speed, so that the anticipation of an accident occurs instantly. The "sin" of the first man: isn't it familiarly the "fall," the ancients thus establishing a direct relation between what they've

> agreed to call original sin and this earthly weight that they considered to be a natural motor of free acceleration of bodies, of their projection, but also of their collision?[145]

A footnote for this passage clarifies what Virilio is communicating poetically: "The man *âdam* comes from the earth, *âdama.* 'Garden' is translated 'paradise' in the Greek version. Eden is a geographical name that can't be tied down to any particular place; it might first have meant 'steppe.' "[146] The point being made, then, is twofold. First, as Adam was fashioned from the soil of creation and animated by the breath of God, Adam's, and by extension Eve's, fearful retreats from the soil of creation after the Fall are analogous to their fearful retreats from each other, their own selves, and God. Second, Adam's and Eve's postures toward the soil of creation are analogous to their postures toward each other, their own selves, and God. The reference to the German romanticist Novalis drives Virilio's poetic-theological point even further because, as Novalis writes, "Our body is part of the world—better put, a member: it expresses the independence, the analogy with the whole—in short, the concept of the microcosm. To this member the whole must correspond: so many senses, so many modes of the universe—the universe a complete analogy of the human being in the body—soul and mind."[147] In short, after the Fall, retreat from the soil is analogous to retreat from others, ourselves, and God; aggression toward the soil is analogous to aggression toward others, ourselves, and God.

In other texts, Virilio's readings of the Fall cause him to rework a traditional Augustinian conception of the biological propagation of sin. For example, Virilio writes cheekily, but also very seriously in *Negative Horizon*, "Man is the passenger of woman, not only at the time of his birth, but also during their sexual relations [. . .]. Paraphrasing Samuel Butler, we could say that the female is the means that the male found to reproduce himself, that is to say, to come to the world."[148] Negatively rendered: fallen and driven men so often use women as vehicles for the prolonging and enhancement of the human race. Positively rendered: God chose the vessel of a woman's womb to come into creation and save creation. Hence, Virilio frequently quotes Saint Hildegard, to entreat: " 'Mankind look at yourself, you have in you the sky and the earth,' the poet Hildegard von Bingen divined, confirming, before [Saint] Francis, the *unison* of measure and excess. She concluded that: 'Man is the enclosure of God's marvels.' Any man is thus not so much the centre of the world as the end of the world. Everywhere he passes the desert spreads."[149] The bodies of fallen Adam, Eve, and the human race are either desolate vehicles or fertile vessels. The desolation or fertility depends on postures toward God, God's creation, the other, and the self. Choose carefully, Virilio warns, because desolate vehicles perpetuate aggressive drives, speed, fear, and violence. Fertile vessels open up to God to receive God's grace and God's marvelous gifts; the most marvelous and gracious gift from God being God's Son, Jesus Christ. In these poetic ways, Virilio reframes a traditional Catholic—particularly an Augustinian—conception of original sin's transmission through biological propagation.[150] From this perspective, the differences between Virilio and Ellul's readings of the Fall, and how they function in their work on technology, becomes clear. Whereas Ellul focuses on the propagation of sin primarily through language and naming (the discursive justification of *technology* and *propaganda* apart from God), Virilio focuses on the propagation of sin through mobile bodies (the start and continuity of *drómos*, the human race).

For this reason, Virilio seems to work with Augustine's accounts of sin and propagation to imply that sinful propagation is determined by fallen *drives*, in contrast to, Ellul's implication that sinful propagation is a *linguistic* concretization of humanity's self-justification of good and evil in separation from God. So, again, in *Negative Horizon*, Virilio demonstrates—more seriously this time—how postlapsarian drives (sins) cause "transfer[s] of violence" in which "man mounted his horse [in] descending from the arms, from the back of woman."[151] Such fallen drives cause "a phenomenon of the metempsychosis of the living, the couple at odds from their origins henceforth from only one body, as in marriage. It is the erotic desire for prosthesis that sets it off in the beginning." Then, "In this instant, the race becomes a higher form than the hunt, *the eruption of the beyond*, of a beyond of physical bodies, territorial and animal, an image of delirium and possession that will, in medieval belief, become the 'diabolical hunt' where the horse takes on an apocalyptic dimension, where the four riders symbolize the end of time and the extermination of history."[152] From this perspective, Virilio's *dromological* studies of fallen humanity's drives, vehicles, and projectiles begin at the Fall, with the coupling, motion, and climax of sexual intercourse. As such, these drives are caused by a fallen desire to possess the other person, reproduce the self, and dominate creation, all in opposition to God. Following this primal drive, the human race frequently becomes, or uses, possessed vehicles that drive it into the "diabolical hunt," which the Book of Revelation unveils so cryptically in the forms of horses and riders.

Returning to *Aesthetics of Disappearance*, Virilio's readings of the origins of the human race (*drómos*) at the Fall cause him to conclude the text with a revealing phenomenological-biographical scene of life during and after the war. He writes, "The over- production of movement implied by war changed the way things looked [to me]; the motor, since it relates directly to the state of paradoxical wakefulness, replaced the causal idea—that was its revolution [double meaning of *perpetual motion* and *revolt*]; *the motor proceeds from the soul*. And so when the playful [Georges] Méliès disguises himself as

Satan of the operetta, doesn't he hit the nail right on the head: the tree of the motor [the engine's drive shaft] is the tree of knowledge (science), corruption of sight is corruption of life!"[153] Georges Méliès was an illusionist and pioneer in modern film direction and editing. He especially was influential in the pioneering of special effect techniques such as jump cuts, splicing, multiple exposures, and reverse sequencing. The film Virilio alludes to in this quote is *The Merry Frolics of Satan*, initially titled *Les Quatre Cents Coups du diable* (*The Four Hundred Tricks of The Devil*) released in 1906. It adapts the Faust myth in which an engineer is deceived by the devil and the devil's accomplices. By the end of the film the engineer is on a train to hell. Once he arrives, he is tied to a spit and spun eternally over flames. Méliès, similar to other pioneers in early photography and cinema such as the Lumière Brothers and Etienne-Jules Marey, often appears in Virilio's analyses of technologies that manipulate light and motion, and in the process, alter peoples' perceptions and the environment.[154] One could say, then, that Virilio reads cinematic *special effects* as the negative obverse of *special revelation*. Beyond the significance of Virilio's reference to Méliès, his description of the "tree of the motor" as being "the tree of knowledge" is especially important to consider. For Virilio, it means that, once Adam and Eve ingest the fruit of the Tree of Knowledge of Good and Evil, it is the Tree of Death, rather than the Tree of Life, which becomes the drive shaft of their souls. Consequently, it is the Tree—or drive shaft—of Death—essentially, Freud's famous theory of the *death drive* articulated in a biblical-theological register—which perpetually drives Adam's and Eve's bodies away from God and God's creation, always mobilizing creation, and consequently, always producing endless material for fast vehicles and violent projectiles. So it is for pre- and postlapsarian bodies, but what about sight and light? Note Virilio's subtle mention of Satan in this quote because he explicitly describes the roles of Satan, sight, and light at the Fall in other texts. For example, in *Negative Horizon*, he writes:

> *It is the last form of anthropocentrism, a new and original form that appeared with the emergence of the second sun that is the motor*, the focus of speed and images with respect to which the "voyeur-voyager" becomes repeatedly *refocused* to the point of being *focused* in the most total inertia since it corresponds finally to the axiality of the geophysical extension comprised in the moment of the gaze: *the motor* [*l'arbe moteur*] [literally: the motor or drive shaft] *becomes the tree of life* [*l'arbre de vie*], *the axis mundi*. Ever exposed to the speed of solar light, we are henceforth overexposed to the light of vehicular speed produced and projected by our machines.[155]

A seductive light (*phōs*) appears at the Tree of Knowledge of Good and Evil to drive humanity away from God and God's creation. By inviting Adam and

Eve to look at the Tree's fruit in a seductive light (*phōs*), the serpent moves (*kīnēsis*) the souls of Adam and Eve to destabilize the bodies of creation (*phusis*), disrupt the senses of humanity, and kick off the human race (*drómos*). So, it is the appearance of this seductive lighting at the Tree of Knowledge of Good and Evil, which causes Virilio to question the West's idealization of reason; and reason, of course, reached peak idealization at the time of the En-light-enment, laying the roads and the tracks for the Industrial Revolution and the modern technologies which followed.[156] As Virilio puts it in *Speed and Politics*, "Reason (as in the *Bible*) is, for the body, a form of death."[157] Proverbs 14:12 and 16:25 affirm this take on human reason: "There is a way that seems right to a man, but its end is the way to death." Applied to Virilio's work on technology, speed, motion, and light, Virilio stresses that the seemingly reasonable way of speed, motion, and light attained through more powerful and further-reaching technologies constitutes a way that will drive us towards death.

While Virilio's writing tends toward a stylized poeticism in comparison to Ellul's more straightforward and lucid prose, Virilio also can be direct about his understanding of how seductive lighting (*phōs*) factors into Genesis 3. For example, in *Ground Zero*, he writes, "To speak [of] the light which comes from the West, and of a world which will not be able to *escape Progress*, is to fall inadvertently into the biblical company of Lucifer (the Bearer of Light), into the tragic irony of Düher's *Melancholia*, and to forget that the Judaeo-Christian story of Genesis is *the story of a scientific suicide*."[158] *Melancholia* is a famous engraving by the German Renaissance artist Albrecht Düher. It depicts a despondent, winged, woman—presumably an angelic personification of melancholy—surrounded by various scientific and technological instruments, which represent numerology, alchemy, geometry, engineering, and so forth. In the upper- to middle-left corner of the print, there is a shining light and a bat-like creature carrying a banner with *Melancholia § I* scripted on it. I take Virilio's reference to the print to be an allusion to the Fall that communicates his skepticism about idealizations of reason and science, beginning at the Renaissance, and developing into the Enlightenment. With quotes such as these, Virilio's point for so much of his writing, which is theologically framed by his readings of the Fall, consequently becomes: watch out for the so-called light (*phōs*) of reason! If it cuts off bodies and senses from the mind, reason can, and often does, fall prey to seductive light—just as Adam and Eve fell when they were seduced by a bad light at the Tree of Knowledge of Good and Evil. Beginning with light (*phōs*) and proceeding to motion (*kīnēsis*) and speed (*drómos*), Virilio shows via his readings of the Fall how peoples' senses of creation (*phusis*) always are mediated by light, motion, and speed. As a consequence, he repeatedly warns, just as Dietrich Bonhoeffer warned, that immediacy is an imposture and an illusion.[159] As much as we may think, we can have an immediate apprehension of ourselves, others, and the world in which we exist, such apprehension is always mediated. The question for Virilio becomes, then, what medium or which mediator will primarily mediate our apprehension of ourselves, others, and the world around us? This question will be explored in a later chapter. For now, it is important to continue with this section's focus on the Fall in Virilio's work on technology.

In *A Landscape of Events*, Virilio more explicitly connects the theological themes examined earlier to the Enlightenment. In this text he writes, the Enlightenment represents a moment in time when, "Westerners grew used to *seeing androids everywhere*, a bit [like] the way Descartes, in denying the power of sensory knowledge, ended up acquiring a somewhat surrealist view of the world." Then, critically, the Fall makes an appearance:

> The android attribution [of Descartes], whether involving pure mechanics or some figuration of the living being, thus offers no Promethean vision of the world but the reverse: a nihilistic coming of its image, a tendency to break up the original unit of beings and things by means of those machines and automata, each of which is, in its own way, a scrap of movement, a piece of intellection, an animal portion, a spare part of time. . . . This dissection being made official at the end of the Age of Enlightenment when the *trunk of the tree of knowledge* (philosophy-physics and technology-theology) splintered and the French revolutionaries, with the baying cult of the "Goddess Reason," declared a sort of deification of the android model, proffered to the crowds as a *substitute faith*, a new state religion.[160]

Beginning with light (*phōs*), and proceeding to motion (*kīnēsis*) and speed (*drómos*), Virilio wants to show how peoples' senses of and thoughts about creation (*phusis*) always are mediated, and not always for the better. The Fall is a primal witness to this truth for Virilio because it was the false enlightenment offered by the serpent, which seduced Adam and Eve to focus only on the fruit on the tree's branch, rather than the entire tree, not to mention the one tree among many in the lush garden offering fruits pleasing to the sight and nourishing for the body (Genesis 1:11–12; 2:9, 15–16). "All that appears appears in light," Virilio often observes in reference to the Apostle Paul, and for that reason, fallen creatures ought to be sensitive to the light (*phōs*), which always mediates their senses.[161] A later chapter draws out further the theological and ethical implications of this understanding of immediacy, mediation, and light in Virilio's writings on technologies. However, here it can be put forward provisionally, that, for Virilio, as well as for the Christian, the true light (*phōs*) of the world does not come from the West. Rather, the true light of the world comes down from above (Incarnation, John 1:9);

rises in the East (Resurrection, Luke 24:1–10); and will return in a flash like lighting from East to West (Eschaton, Matthew 24:27–28). All of creation will be crossed—north, east, south, west—by the resurrected and returned Jesus Christ. It is this light that ought to mediate a Christian's and church's speeds (*drómos*) and motions (*kīnēsis*) within the environments (*phusis*) that *epistēmē*, *poiēsis*, and *tékhnē* are situated.

The next chapter shows how postlapsarian social power, spiritual powers, and technology dangerously mix together in Ellul's writing to raise the already significant stakes of his ethics of technology. For Ellul, social power and spiritual powers are consequences of the Fall that he examines from historical, sociological, theological, and ethical perspectives throughout his writings on technology. In contrast, Virilio uses history, philosophy, and phenomenology to expose the connections between postlapsarian power, powers, and technology. For this reason, making room for a consideration of both authors' understanding of power, powers, and technology will add substantial weight to the ethics highlighted and developed in the sixth and final chapter.

NOTES

1. Ellul, *Technological Society*, 23.
2. Ibid.
3. Ibid., 23–24.
4. Ibid.
5. Ibid., xxx–xxxi. Apart from Ellul's dialectical method accounting for this decision he writes in the introduction that the "Probable development I describe [in *Technological Society*] might be forestalled by the emergence of new phenomena." The first possible phenomenon is a general war, which causes enough destruction that the conditions for survival prevent society from becoming fully technological; the second is an increasing number of people become aware of the threat technology poses to people's personal and spiritual lives and they respond with resistance; and, the third is the possibility that God decides to intervene in history to influence technology's direction. Ellul concludes that none of these possibilities can be considered from a sociological perspective—especially the last two—but, of course, in other works, he does consider the last two possibilities from a theological perspective.
6. *Oxford English Dictionary Online*, "Homo faber, n." March 2019, accessed March 5, 2019, http:// www.oed.com/view/Entry/272606?redirectedFrom=homo+faber&. This chapter and later chapters will elucidate on the significance of Greco-Roman antiquity for Ellul's dialectical writings on technologies so it is noteworthy that Appius Claudius Caescus, the starting point for *homo faber*, lived precisely within this time period.
7. Ellul, *Technological Society*, 24.
8. Ellul, *New Demons*, 91.

9. Karl Marx and Frederick Engels, *Marx/Engels Selected Works, Volume One*, trans. Samuel Moore and Frederick Engels (Moscow: Progress Publishers, 1969), 17.

10. Ellul, *Technological Society*, 27.

11. Ellul returns to this theme in *Hope in the Time of Abandonment*, specifically on pages 38–42, in a section titled "The Age of Magicians." He writes about the paradox of a modern "recourse to magic," which signifies an attempt "to recreate the sacred" within a technological society that supposedly is, but actually is not, secular. His basic point is the same, however, in that such recourse is a "purely social phenomenon."

12. Ellul, *Meaning of the City*, 8.

13. Karl Marx and Frederick Engels, *Marx/Engels Selected Works, Volume One*, trans. Samuel Moore and Frederick Engels (Moscow: Progress Publishers, 1969), 17.

14. For example, see Ellul, *Ethics of Freedom*, 23–25 in which a connection is formed between Marx's concept of alienation and Ellul's understanding of sin.

15. Ellul, *Technological Society*, 27–28. I am uncomfortable with Ellul's interchangeable use of the terms *East* and *Orient*. However, given he chose to describe the history of *technique* as such in this passage, and given he resists—demythologizes, as it were—a typically orientalist reading of the East as mystical and non-pragmatic, I have chosen to leave the terminology as is.

16. Ibid., 7. Ellul writes, "The first techniques of Hellenistic civilization were Oriental; they were not derived from Greek science. Thus, historically speaking, the relationship between science and technique ought to be reversed."

17. Ibid., 33.

18. Ibid.

19. One might also accuse Ellul of glorifying the West, and it is possible to read him in these and other instances as implying that Rome is the center of civilization. It has already been shown that Ellul admired certain aspects of Greco-Roman Antiquity—particularly Roman law—but Ellul is also obviously critical of the West, to the extent that, he felt the need to write a book like *Betrayal of The West* to balance his critique of Western civilization. He writes in the prologue (page vii) of this text: "I, who have attacked the technical society and its scientific rationality, feel obliged to show that there is also a very different side to the West. The West represents values for which there is no substitute. The end of the West today would mean the end of any possible civilization." While I appreciate Ellul's attempt to correct over-readings of his criticisms of the West, his attempted corrective at times can be . . . disagreeable.

20. Ellul, *Meaning of the City*, 3. All preceding quotes in this paragraph are from this page. See also Jacques Ellul, *La Genèse aujourd'hui* (Paris: Editions de L'AREFPPI, 1987), 111–112 for a similar discussion regarding the significance of Eden being set in the Orient, Cadesh, and the East.

21. Ibid.

22. Ibid.

23. Ibid.

24. Ellul actually mentions Augustine's "City of God" on page 160 of *Meaning of the City*; he also mentions "two directions," which are "basic to the history of the city. One originates in Cain, the other in Eden, and they converge finally in the New

Jerusalem" on page 163; and mentions of Augustine's comments on "those who are predestined" on page 198.

25. Augustine, *City of God*, XV.15.2. Elsewhere, in *Presence in the Modern World*, 27 Ellul writes, "The first condition [of Christian faith in the modern world] is a well-known truth, but not sufficiently understood in its reality: *Christians belong to two cities*." This is evidence that Ellul was well-acquainted with Augustine's theology and applied it to his own understanding of Christianity in the modern world.

26. Martin Luther, *Commentary on Genesis* Vol. 1, IV.1.2a.

27. Ellul, *Meaning of the City*, 6.

28. Ibid., 6.

29. Peter N. Stearns, "Pinpointing the Industrial Revolution: Features, Times, and Place," in *Debating the Industrial Revolution* (New York: Bloomsbury Publishing, 2015), 13–34 demonstrates the degree to which what commonly is referred to as "The Industrial Revolution" is itself a *myth*. Scholars cannot seem to agree upon when and how this "revolution" occurred and for how long it lasted. This, it seems, proves all the more Ellul's point about sociological *myths* and their institution of sacred orders in society. Apart from historians, most people have a vague understanding of the Industrial Revolution and its importance in modern history, but few can explain directly or fully what it is and when it happened beyond mentioning that it represents a time when "machines appeared" and that those machines "changed industry." But this is enough, it seems, to foster a mythical fascination with modern machines.

30. Ellul, *Meaning of the City*, 6. Emphasis added and term included for clarification.

31. Ellul, *La Pensée Marxiste*, 46.

32. Ellul, *Judgment of Jonah*, 80. As it is for Cain, so it is for Jonah. Ellul writes, "When Cain, in face of the first act of God's good will, the acceptance of Abel's sacrifice, becomes angry and is full of envy and murder, God says to him too: 'Why are you angry?' and at that moment Cain decides to become a murderer, so that it is not for nothing the question is repeated here. Jonah, too, is angered by God's benevolence and he demands the execution of the threat, the slaying of the people of Nineveh. Hence he, too, becomes fratricide. He even asks God to take part in the murder. He becomes Cain, for he is in truth the brother of these people in his revolt, his condemnation, and the pardon he receives."

33. Ellul, *Meaning of the City*, 6.

34. Ellul, *Technological Society*, 28.

35. Ellul, *Théologie et Technique*, 116.

36. Ellul, *Humiliation of the Word*, 206–242 demonstrates at length Ellul's understanding of the differences between and relationships of "reality" and "truth."

37. Ellul, *False Presence of the Kingdom*, 202–205.

38. Ellul, *Technological Society*, 23–24.

39. Ibid.

40. Ibid., 24.

41. Ibid.

42. Ibid., 27.

43. Bringing this text into the discussion is warranted here because of Ellul's Kierkegaardian opening to *Reason for Being* on page 4 of this text: "Whereas *The Presence of the Kingdom* formed the general introduction to all of what I wanted to write, [my commentary on] Ecclesiastes will be the last word." Like *Presence in the Modern World, Reason for Being* is not a self-contained work in Ellul's oeuvre but a hermeneutically important one for reading all of Ellul's writing. It is likewise important that *technique*, among other key themes in Ellul's work, receives substantial commentary in *Reason for Being*. Pages 91–93, 152–153, and 301–303 are examples.

44. Ellul, *Reason for Being*, 58–60 features the entire two-page commentary on Abel's name of which this quote is a part.

45. Ibid., 52.

46. Ellul, *Technological Society*, 27.

47. Marx, *Manifesto for the Communist Party*, 16.

48. Ibid.

49. Ibid.

50. Ellul, *Technological Society*, 24.

51. Ibid., 23.

52. Ibid., 26.

53. Ibid., 24.

54. Ibid.

55. Mediatory sacrifice has ties to paganism and certain versions are described anthropologically, sociologically, and historically as "magical *technique*." Ellul's engagement with René Girard in other texts communicates that such language and analysis is familiar to Ellul, and at the same time, that Ellul is clear on the difference between the mediatory sacrifices of paganism and the mediatory sacrifice of Jesus Christ. For example, see Jacques Ellul, *Subversion of Christianity*, 159–160: "Girard is quite right when he shows how basic sacrifice is to humanity. There can be no accepted life or social relation without sacrifice. But gracious grace rejects the validity of all human sacrifice" because "by his life and death, [Jesus] shows that those who cannot be justified before God are in fact justified by grace and the love of God."

56. Ellul, *Reason for Being*, 58.

57. Ellul, *Meaning of the City*, 8.

58. Ibid.

59. Ibid.

60. Ellul, *Technological Society*, 23.

61. Ellul, *Meaning of the City*, 5–6. Genesis 4:2 reads "Abel was a keeper of sheep, and Cain a tiller of the ground."

62. Ibid., 12.

63. Ibid., 5–6. "Cain dedicates a new world: 'Enoch,' as opposed to *Reshith* in Genesis 1:1. Inauguration, as opposed to creation."

64. Ibid., 14. Ellul is influenced by but does not fully accept Marx's understanding of work as a constitution of the human. A passage like this in *Meaning of the City* demonstrates how Ellul sees *technique*, and by extension work, as a reflection of the human but not a constitution of the human. It is a nuanced existential distinction being made here, but it is an important one because it establishes a difference between

saying people are defined by their work and saying that something of consequence about people is revealed in their work. Ellul favors the latter reading and repeatedly critiques commonplace conceptions of work as blessed and virtuous. Also on page 169 of *Meaning of the City* Ellul responds directly to Marxists with the suggestion that "the city is an almost indistinguishable mixture of spiritual work and man's work" and "man is clearly not the one to change that inner spirit as he wishes—we see every day that he cannot change even the outer face." See *Reason for Being*, 93–95; *Ethics of Freedom*, 128–129; and *Théologie et Technique*, 156–157 for other examples of delivering his critique of work from a theological basis. In general, he repeatedly refers to Genesis 3:15 to point out that, at least after the Fall, work is a "curse." I see problems with this aspect of the theological bases for Ellul's arguments here, because, as he tries to resist idealizations of work and human works after the Fall, he argues in texts like *False Presence of The Kingdom*, 28–29, to critique modern justifications for work and technologies: "[In Genesis 1 and 2] we are told that the creation was completed, that it was perfect and that there was nothing to add to it." Yes, Genesis 1 and 2 communicate God's completion of creation, but both texts use the words "good" and "very good" to describe God's views of speaking about "finished" creation, and also, they describe God finishing "the work that he had done" and say that God "rested on the seventh day from all the work he had done." These points are not made to justify fallen work, works, or technology, but to point out that Ellul either intentionally misrepresents or unintentionally misreads Genesis 1 and 2, thereby causing him at times to deploy theologically questionable readings of work and pre- and postlapsarian creation.

65. Ibid., 14. I add emphasis to "the cradle of urban civilization" to highlight another passage in which Ellul states that the East is the cradle for technical action.

66. Ellul, *Technological Society*, 27–28.

67. Ellul, *Meaning of the City*, 14.

68. Ellul, *New Demons*, 51–52.

69. Ellul, *Meaning of the City*, 14–15.

70. Ibid.

71. Ellul, *Technological Society*, 63–64.

72. Ibid.

73. There quite a lot of discussion in Ellul scholarship about whether or not there is a qualitative difference between classic *tékhnē* and modern *technique* and whether or not Ellul thinks *technique* is a novel phenomenon unique to modernity. My understanding and engagement with this discussion is summarized here but described at length in my PhD dissertation. For those interested in further details about this aspect of Ellul's work and Ellul scholarship, see Michael Morelli, *Either Technological Pessimism or Proleptic Christian Resistance: A Constructive Reading of Jacques Ellul's Dialectical Writings on Technology with Paul Virilio as an Evocative Interlocutor*. University of Aberdeen, Aberdeen, PhD Dissertation, 2019. https://ethos.bl.uk/OrderDetails.do?uin=uk.bl.ethos.787402, particularly chapter 3. Also, I think Virilio agrees with Ellul that there is a certain qualitative novelty to modern *technique*. For example, see Virilio's discussion of the "three phases of military knowledge" in *Pure War*, 30–31. There, he says "[in the third phase of military

knowledge] we have the great surprise, no longer technical but scientific, a surprise of another kind: the advent of the nuclear bomb. It's no longer a *quantitative* problem that surprises the military staff, and thus the states; now it's a *qualitative* problem: the ultimate weapon. Logistics takes over." Emphasis added. Basically, once the quantitative development of technology throughout history leads to point when and where the entire world could be obliterated by an atomic bomb, there is, in fact, a qualitative shift that comes with modern technology, even if the human spirit and will which drives technological innovation throughout history is qualitatively the same.

74. Ellul, *Meaning of the City*, 113.

75. See the discussion in the previous chapter's section "Theological Myth" regarding the important passages from Ellul, *New Demons*, 112–113.

76. Ellul, *Meaning of the City*, 15–16.

77. As it is not pertinent to the focus of this chapter or book, I will not be outlining Ellul's mythological return to zero for *propaganda* here. However, commentary on this aspect of Ellul's work is featured in my dissertation, which can be viewed at Michael Morelli, *Either Technological Pessimism or Proleptic Christian Resistance: A Constructive Reading of Jacques Ellul's Dialectical Writings on Technology with Paul Virilio as an Evocative Interlocutor*. University of Aberdeen, Aberdeen, PhD Dissertation, 2019. https://ethos.bl.uk/OrderDetails.do?uin=uk.bl.ethos.787402. One aspect of this commentary that may be of interest to readers is Ellul's location of the historical origins of "false news" (i.e. *fake news*) in his untranslated *Histoire de la Propagande*. See Jacques Ellul, *Histoire de la Propagande* (Paris: Presses Universitaires de France, 1967), 11 for relevant content on this topic.

78. Ellul's reading of the Fall shapes his work in significant ways, so I unfortunately cannot cover the entirety of this aspect of his work here. Instead, I am focusing on the elements that relate to the discussion at hand. For other scholars who engage well with Ellul's reading and application of the Fall in his work, see Goddard, *Living The Word*, 75; 77; 82–83; 145–147; 151; and 155; Daniel Clendenin, *Theological Method in Jacques Ellul* (New York: University Press of America, 1987), 65–69; 77; and 81–84; Lovekin, *Technique, Discourse, and Consciousness*, 20; 54; 70–72; 100–101; and 164; Marva Dawn, *The Concept of "The Principalities and Powers" in Jacques Ellul* (Notre Dame: PhD Dissertation, 1992), 93–94; 114; 116–117; 170; and 187–188; Temple, *The Task of Jacques Ellul*, 318–320; 330; 336; 340–343; and 347–346; and, Brian Brock, "Culture as Flight From God," *Fall Narratives*, eds. Zohar Hadromi-Allouce and Áine Larkin (London: Routledge, 2017), 37; 40–41; and 44–46.

79. Ellul, *Humiliation of the Word*, 67–68.

80. Ibid.

81. Ibid.

82. Ibid.

83. Ibid.

84. Ibid. Ellul's emphasis.

85. As mentioned earlier, see Morelli, "What's in a Name? Jacques Ellul's Reading of Naming in Genesis 1–3," *Jacques Ellul and the The Bible*, 82–95 for further commentary on Ellul's reading of naming in Genesis 1–3. Suffice it to say here,

I explore in this piece what it means for Ellul that Adam names his wife "woman" in the form of a poem or song before the Fall (Genesis 2:21–22) and then renames her "Eve" after the Fall (Genesis 3:20) in a way that seems to constrict her to the function of childbearing.

86. Ellul, *To Will and to Do*, 5.

87. Ibid., 6.

88. Ibid.

89. Ibid., 10–11. Ellul's emphasis.

90. Similar readings of the Fall appear in many other texts of Ellul's. See, for example, Jacques Ellul, *Theological Foundation of Law*, trans. Marguerite Wiser (New York: Seabury Press, 1960), 83–84 and *Presence in the Modern World*, 48.

91. Ibid., 12. The theological importance of "nudity" for Ellul's work is examined later.

92. Ellul, *Humiliation of the Word*, 67–68.

93. Ellul, *Presence in the Modern World*, 48. Another possible rendering of "it is obvious" in the original French is "it is evidence," in the sense of *evident* and *observable*. Which is to say, Eve observes with her eyes and senses, and concludes in her mind, that the fruit is worth eating after dialogue with the serpent. So, "obvious" may be rendered better as "evidence" in this passage. The latter translation communicates in clearer terms how what is self-evident is justified by people's problematic observations of material phenomena, rather than, the material phenomena itself being "obviously" good.

94. Ellul, *Technological Society*, 79–80. John 14:6.

95. Ellul, *Humiliation of the Word*, 20–21 and 206–207.

96. Ibid., 93. It is noteworthy that Ellul opens the section with "we must constantly return, however, to the Fall."

97. Ibid., 94–95. See also, Ellul, *To Will and to Do*, 11–13 and Ellul, *Théologie et Technique*, 152–153. Again, nakedness and clothing before and after the Fall appears, and as such, it will be examined at a later point.

98. Ellul, *Meaning of the City*, 148.

99. Ellul, *Théologie et Technique*, 154.

100. Ibid., 158.

101. Ellul, *Meaning of the City*, 5–6; 9; 15–17; and 18–19. Ellul, *Humiliation of the Word*, 97–98. Or, for a lengthier commentary on The Tower of Babel, "Religion and Revelation," *Living Faith*, 126–156.

102. Ellul, *Meaning of the City*, 14. Ellul, *Technological Society*, 27–28. To avoid belaboring the point that the Fall is Ellul's return to zero for *technology*, here are further references that affirm and build upon what is cited and commented on in this and the previous section. Ellul, *Théologie et Technique*, 91–92; 99–100; 110–111; 144–145; 147–149; 152–154; 155–159; 163–164; 184–185 200–201; 236–237; 263–264; and 334–335.

103. Ellul, *Technological Society*, 144–145.

104. Ibid., 40.

105. Ellul, *To Will and to Do*, 5.

106. Ellul, *False Presence of The Kingdom*, 30 and 26–28 for further context, including comments on how Genesis 1, 2, and 9, as well as Psalm 8 factor into his account of justification and technologies.

107. Ellul, *Hope in Time of Abandonment*, 152.

108. Ibid.

109. Ibid. For other examples see Ellul, *Théologie et Technique*, 268–269 and 271. Ellul, *Reason for Being*, 92–95; 98–99; and 104.

110. Ellul, *To Will and to Do*, 190.

111. Ellul, *Ethics of Freedom*, 143. 143–145 for further context. Ellul goes on to say on page 145: "The decisive and constantly repeated act of man is to separate [God's] word from the one who speaks it and to try to make [God's word] his own. Adam did this. Faced by the one commandment of God, he isolated it under the serpent's influence. He set God aside and controlled the commandment by giving it another point and meaning. Finally he made the commandment into a word of his own by himself saying what is good and evil."

112. Ellul, *Humiliation of the Word*, 67–68.

113. Virilio, *City of Panic*, 12. Virilio's emphasis. For other examples, see Virilio, *Negative Horizon*, 42; 115; 117; 121; and 147; Virilio, *The Original Accident*, 5–6; Virilio, *Grey Ecology*, 50; and, Virilio, *The Great Accelerator*, 11.

114. Virilio and Lotringer, *Pure War*, 34–36.

115. Virilio and Armitage, *Virilio Live*, 83. See also Virilio, *University of Disaster*, 36.

116. Virilio, *Negative Horizon*, 69.

117. Virilio does not provide a citation for his quote of Sombart in the *Negative Horizon* passage, but it likely came from one of the three volumes of *Modern Capitalism*, Sombart's most well-known work. For more on Sombart, see Hugo Reinert and Erik Reinert, "Creative Destruction in Economics: Nietzsche, Sombart, and Schumpeter," *The European Heritage in Economics and the Social Sciences* 3 (2006): 55–85. It perhaps goes without saying, but it may be worth mentioning in order to clarify: Virilio's Sombart reference does not reflect his agreement with Sombart's antisemitism. Just the opposite, in fact. Virilio often uses the quotes of immoral, depraved figures in history to expose its darker and bloodier aspects—and, through such exposure, shed light on similar or novel forms of immorality and depravity present in our world today.

118. For example, see Ellul, *Technological Society*, 21; 43; 53; 87; 110; 343; and, 421–422

119. Ibid., 421–422 is one such example.

120. Virilio and Armitage, *Virilio Live*, 26.

121. Virilio, *Crespuscular Dawn*, 102–103. See also Virilio, *Politics of the Very Worst*, 14; Virilio, *City of Panic*, 126–127; and Virilio, *Negative Horizion*, 62.

122. Virilio, *Ground Zero*, 23.

123. Virilio, *Speed and Politics*, 111. Virilio's emphases.

124. *Meaning of The City* was published in English in 1970 and published in French as *Sans feu ni lieu: Signification biblique de la Grande Ville* in 1975. Ellul also published writings on Genesis 4, Cain, *technique*, and the city as early as the forties

and fifties. For example, see Ellul, *Theological Foundation of Law*, 86–87; 105–106; Jacques Ellul, "La Bible et la ville," *Foi et Vie* 48, no. 1 (janvier-février 1950): 4–19; and Jacques Ellul, "Cain, the Theologian of 1969." *Katallagete* (1969): 4–7. Virilio could have encountered any or all of these publications before beginning work on and publishing *Speed and Politics* in 1977.

125. Virilio, *University of Disaster,* 134–135.

126. Benjamin, *Illuminations*, 263–264.

127. Benjamin, *Illuminations*, 263–264.

128. Virilio, *L'insécurité du territoire*, 163–178. Other examples are Virilio and Lotringer, *Accident of Art*, 35; Virilio, *City of Panic*, 129; Virilio, *University of Disaster*, 124 and 134–135; and Virilio and Brausch, *Winter's Journey*, 117. Light's thematic prominence in many of these passages is noteworthy.

129. Virilio, *L'insécurité du territoire*, 167–168. Here Virilio takes aim at Cartesian rationality and what he describes as a morality imposed secondarily on the physical environment via detached investigation and calculation. Virilio argues that what has come to be understood as "natural" in regard to "natural law" has actually been domesticated and attacked by a form of metaphysics, which does not take into account the actual physics (*phusis*) of the environment, which is the only substance out of which, Virilio thinks, any true meaning could arise or grow (*phúein*). Of course, technologies only add to the problem because they tend to detach people from the world, and as a consequence, technologies cause people to think about the physical environment—from which meaning might grow—in increasingly abstract ways.

130. Virilio and Lotringer, *Accident of Art*, 35. "[I'm paraphrasing] mainly the *Old Testament*, especially psalm 18: language without speech. God speaks without speech before the Prophets. Before Israel. God speaks in Creation: the sun speaks to the night, the night to the day. Clearly, language without speech is the language of Creation. And Creation is nature. It's the sun, the beauty of the sea, etc., but it is also the creation of humanity. It's the silence of painting, the silence of the impressionists."

131. Paul Virilio, *Strategy of Deception*, 21. Virilio writes, "After the eye of God pursuing Cain into the tomb, we now have the *eye of Humanity* skimming over the oceans and continents in search of criminals. One gets an idea, then, of the *ethical* dimension of the **global information dominance** programme, the attributes of which are indeed those of the *divine*" (Virilio's emphases).

132. Virilio, *Speed and Politics*, 111. Virilio's emphases.

133. Virilio, *Speed and Politics*, 32–33 and 40; Virilio and Lotringer, *Pure War*, 18–19; Virilio and Petit, *Politics of the Very Worst*, 39 and 79; Virilio, *Desert Screen*, 4; Virilio, *City of Panic*, 124; and Virilio and Brausch, *A Winter's Journey*, 53. Ellul says much the same, in, for example, *Meaning of the City*, 13: "Spirit of might, conquest, construction of cities. The city is thus tied in with the will of man whose characteristics we have already stated and which we will meet again. The city is now a center from which war is waged. Urban civilization is warring civilization. Conqueror and builder are no longer distinct. What world could better demonstrate the parallel between urban civilization and warring civilization than our own, a world where the city and war have become two of the poles around which the entire economic, social,

and political life of our time move." But, of course, Virilio is more preoccupied with war in his writing than Ellul is.

134. It is worth noting that Virilio forms symbolic connections in *Speed and Politics*, on 85 and 130 between Cain the "warrior" and "perverted priest" and Teilhard de Chardin, who Virilio calls a "military monk." So, it seems Virilio thinks Cain is a figure for Chardin and vice versa, and that symbolizes both the perverting force of the military structure and warrior function. Virilio quotes Chardin, who wrote in 1924, that "we still need mightier and mightier cannons, bigger and bigger battleships, to materialize *our aggression on the world*," and wrote in 1917, "War is an *organic phenomenon of anthropogenesis* that Christianity cannot suppress any more that it can do away with death." Virilio quotes and describes Chardin in a similar light in other texts such as Virilio, *Ground Zero*, 78. This implies that Virilio and Ellul were unified in their skepticism about thinkers such as Chardin, a contemporaneous French Christian, who was much more optimistic about the development of technologies than Virilio and Ellul were. See, for example, Ellul, *False Presence in the Modern World*, 208–209 for one of the many texts in which Ellul similarly criticizes Chardin. Ellul writes, "[Today] we are possessed with more powerful means, and we cover the entire planet. It is the establishing of a total world" and "it is the scandal and the appalling heresy of Teilhard de Chardin to pretend to bless this totalism of the world in the name of Christianity; but that cannot be done, as is evident from his writings, without leaving to one side the incarnation of Christ in the person of Jesus." Also, Bernard Charbonneau wrote a book titled *Teilhard de Chardin, Prophet of a Totalitarian Age*, which Ellul probably read and would have discussed at length with Charbonneau.

135. Virilio, *Speed and Politics*, 111.

136. Romans 8:22–24. Also Luke 19:40.

137. Ellul, *Humiliation of the Word*, 97.

138. Ibid.

139. See, for example, Virilio, *Aesthetics of Disappearance*, 86. The corresponding footnote on page 126 reads, "The serpent serving as a mask to a hostile being in which Wisdom, then Christianity, wanted to recognize Satan. 'The woman saw that the fruit of the tree of knowledge was good to eat and pleasing to look upon.' " It suggests thematic connections with Ellul, *Humiliation of the Word*, 93 and 94 when Ellul writes, "This sight refers to the truth, appropriately enough! In Genesis 3 we read that 'the woman *saw* that the tree was good for food, and that it was a delight to the *eyes*, and that the tree was desired to make one wise' (Genesis 3:6)" and " '*Evidently*,' she sees. Evidence becomes the sign of certainty and conviction" (Ellul's emphases). These textual similarities do not confirm completely that Virilio had Ellul's writing in mind when he developed his own reading of the Fall because, similar to *Speed and Politics*, Virilio also does not cite Ellul in *Aesthetics of Disappearance*. But Virilio did read *Humiliation of the Word* because he cites and engages with it in *Open Sky*. So, there could be an influence of *Humiliation of the Word* on *Aesthetics of Disappearance*.

140. Virilio, *Aesthetics of Disappearance*, 86–87 and 113.

141. Ibid., 86. This reading of the Fall helps explain in theological terms the book's title; why it describes from various perspectives "the cycle of humanity sworn

less to death than disappearance"; and, why Virilio prefaces the book with a quote from 1 Corinthians 7:31: "The world as we see it is passing."

142. Ibid.

143. Virilio does not write at length about prelapsarian creation but what he does write implies a traditional view of God as an eternal creator who is copresent to creation before and after the Fall. See, for example, Virilio, *A Landscape of Events*, x. He quotes Ecclesiasticus 39:19–20. See also Virilio, *Lost Dimension*, 109 in which Virilio quotes Dietrich Bonhoeffer, *Creation and Fall*, ed. John W. De Grucy, trans. Douglas Stephen Bax (Minneapolis: Fortress Press, 1997), 38–39 to describe and philosophically explore the differences between pre- and post-lapsarian creation. The Bonhoeffer quote relates to exegesis of Genesis 1:1–2, and Bax's more accurate translation of what Bonhoeffer writes is, "The creator enormously increases the power of the creation, by giving to creation its own being as that which has form. In this form creation exists over against God in a new way, and in existing over against God it wholly belongs to God."

144. Virilio, *Negative Horizon*, 42–43.

145. Virilio, *Aesthetics of Disappearance*, 87. Virilio's emphasis. Note how "accidents" appear in this passage. The thematic prominence of "accidents" in Virilio's writing is examined in the next chapter.

146. Ibid., 126.

147. Novalis, *Vorarbeiten*, fr. 483, in *Werke*, ii 423. Quoted from Nicholas Halmi, *The Genealogy of the Romantic Symbol* (Oxford: Oxford University Press, 2007), 81.

148. Virilio, *Negative Horizon*, 37.

149. Virilio, *City of Panic*, 131–132. Virilio's emphasis. Also, Virilio, *Politics of the Very Worst*, 88; Virilio and Armitage, *Virilio Live*, 20–21; and Virilio and Lotringer, *Crepuscular Dawn*, 157–158.

150. Augustine, *City of God*, II.XIV, 14–28.

151. Virilio, *Negative Horizon*, 38 and 40.

152. Ibid., 43–44. Revelation 6:1–8. Virilio's emphasis.

153. Virilio, *Aesthetics of Disappearance*, 113. Virilio's emphasis.

154. See, for example, Virilio, *War and Cinema*, specifically the chapters "The 'Fern Andra' Cinema" 66–76 and "A Travelling Shot over Eighty Years," 85–111.

155. Virilio, *Negative Horizon*, 116. Virilio's emphasis.

156. Given Virilio often calls the Industrial Revolution *the industrial transportation revolution*, this section and the previous has established the theological foundations for such descriptors. For examples, see Virilio, *City of Panic*, 12; Virilio, *Negative Horizon*, 117 and 147; Virilio, *The Original Accident*, 5–6; Virilio, *Grey Ecology*, 50; Virilio, *The Great Accelerator*, 11. Also, he often plays with neoplatonic images of light and reason in texts such as *A Landscape of Events*. See, for example "The Big Night" 1–11 and "The Near-Death Experience" 32–36. He also tends to play with traditional theological approaches to reason, wisdom, science, and ethics in texts like *Open Sky*, particularly "Eye Lust" 89–102. He opens the chapter with the questions: "What, in fact, is the true tree? The one perceived in a pause, every detail of which can be visually itemized, every branch and leaf; or the one glimpsed flashing past in the stroboscopic unfolding of the car windscreen, or else through the strange

skylight of television?" Compare this to Book V§7 of Augustine's *Confessions*: "In the same way someone who knows how to take care of a tree and gives thanks to you [God] for the enjoyment of it, even though he has no idea how many cubits tall it is or how wide its spread, is better than a person who measures the tree and counts all its branches, but does not keep it properly or recognize its maker and love him." Each text and thinker in their own ways are alluding to and applying 1 Kings 4:29–34, in which Solomon, who is given wisdom by God and able to discern the various species of creation, still falls prey to the worship of false gods. Special thanks to Andrew Errington for mentioning the 1 Kings passage and its significance for wisdom and practical reason within the Christian tradition.

157. Virilio, *Speed and Politics*, 108.

158. Virilio, *Ground Zero*, 85. Author's emphases.

159. Virilio, *War and Cinema*, 58–59 and 81–82; *Lost Dimension*, 127; *Negative Horizon*, 182; *Ground Zero*, 40; and *University of Disaster*, 59. The quote is from Dietrich Bonhoeffer, *Discipleship*, trans. Green and Krauss (Minneapolis: Fortress Press, 2003), 94–95 and the importance of this for Virilio's philosophy, theology, ethics, and aesthetics is examined in the final chapter.

160. Virilio, *A Landscape of Events*, 33. The reference is from René Descartes, *Meditations*, trans. Elizabeth S. Haldane and G. Ross (Chicago: Encyclopaedia Britannica, 1952), Meditation 11, p. 81. Descartes describes looking out a window, and writes, "What do I see from the window but hats and coats which may cover automatic machines? Yet I judge these to be men."

161. Virilio, *The Vision Machine*, 33; Virilio, *Negative Horizon*, 115. The reference is to Ephesians 5:7–14.

Chapter 5

Power, Powers, and Technology

Just as readings of the immediate consequences of the Fall critically frame how Ellul thinks and writes about technology, so does his understanding of postlapsarian social power and spiritual powers. The language of The Spirit of Power and The Spirit of Lies appear throughout Ellul's dialectical writings, animating *technique*, *technology*, and *propaganda*, and it augments his theology and ethics. These powers instantiate a mysterious spiritual convergence wherein individuals and institutions use technology due to an idolatrous desire for power, but in the process, are used by technologies in damaging ways. Virilio's understanding of power(s) and technology is similar to Ellul's, but he differs in the way he exposes the power(s) manifested by technology, offering a complementary counterpoint to Ellul's method. As a whole, this chapter adds weight to both authors' ethics as they are examined and developed in chapter 6.

PRINCIPALITIES AND POWERS

Ellul's understanding of social power and spiritual powers evolve from more to less substantialist characterizations. Key passages show that The Spirit of Power appears early on in his work and retains its prominence throughout his dialectical writings on *technique*, *technology*, and *propaganda*. As early as *Theological Foundation of Law*, Ellul writes in reference to the *technical* phase of law ushered in by the mutation of classic *tékhnē* into modern *technique*, "The opposite of [appropriate] judgment is, in fact, the spirit of power, which the Beast of the Sea represents for the political and juridical realm."[1] Passages like these are prefaced methodologically by early articles where Ellul writes, in 1946, about providing an "inventory of [the] forces," which "condition the life of our time."[2] As Ellul worked on his plan for a fifty-year writing project during World War II, he decided a significant part of his project would be to "discern the signs of the times" (Matthew 16:3 and Luke 12:56), with particular attention given to work, which would "test the spirits" (1 John 4:1), in an effort to heed the call to be "transformed by the renewing of your minds, so that you may discern what is the will of God" (Romans 12:2).[3] Ellul repeatedly appeals to Ephesians 6:12 to remind readers that the first battle of the church is not a fight with flesh and blood but combat with the principalities and powers of a fallen world.[4] These aspects of Ellul's writing plan make sense contextually speaking. As described in the first chapter, his experiences leading up to, during, and after World War II meant he experienced firsthand the horrible social power and spiritual powers, which animated the first total war in history.[5] So, discernment of and prayer in combat with social power and spiritual powers not only was a critical aspect of Ellul's dialectical writings, it also was a critical aspect of his life as a disciple who lived during terrible moments in history.[6]

In terms of other Ellul scholarship in this area of his work, Marva Dawn's power(s) hermeneutic for Ellul's dialectical writings is especially insightful. For this reason, aspects of her power(s) hermeneutic are a departure point for sections of this chapter, which examine Ellul's work on social power and spiritual powers. Instead of repeating Dawn's excellent work, this section and the next will advance Dawn's broader reading of power(s) in Ellul's dialectical writings into new, finer grained, territory by focusing on the two powers, which animate *technique*, *technology*, and *propaganda*: The Spirit of Power and The Spirit of Lies.[7]

Key passages in *Ethics of Freedom* establish the importance of social power and spiritual powers in Ellul's dialectical writings on technologies. Ellul writes in this text, after a discussion of spiritual powers read through the frame of Ephesians 6:12, "To me the powers seem to be able to transform a natural, social, and intellectual, or economic reality into a force which man has no ability either to resist or to control." He uses the nation-state and money as examples, observing "if [these structures] are viewed simply in their concrete reality and natural dimension, and at the level of their function, it is impossible to understand their power, to measure their zone of action, or to grasp their true realities."[8] Consequently, "The disproportion noted [here] leads me to the unavoidable conclusion that another power intervenes and indwells and uses political power, thus giving it a range and force that it does not have itself. The same is true of money. The question is identical."[9] Finally, and most critically for Ellul's understanding of *technique*, *technology*, and *propaganda*, he writes on the next page "as I see it, money is what it is only because Mammon indwells it. We are in the presence of an economic phenomenon which is quickened by a power that destroys man, that man himself does not create, and that does not issue from the heart of man. These two examples of [the state and money] suggest others. Thus one might ask whether technology is not also one of these powers. The answer seems to be simple enough."[10] A remarkably truncated answer considering how much Ellul wrote about technology!

Given the place of *Ethics of Freedom* in Ellul's oeuvre—chronologically, the first two of the three volumes sit approximately at the middle—the aforementioned passages balance Ellul's more and less substantialist accounts of the principalities and powers as they develop over time.[11] Basically, for Ellul, social power dialectically interacts with spiritual powers to produce a power neither could produce independently. Then, for the individuals and institutions who enter this power(s) dialectic due to an idolatrous desire for power, they eventually are used and damaged by the spiritual powers at work. Returning to what Ellul says about power(s) and technology in *Ethics of Freedom*, then, it is clear that, despite a performative truncation in the previous passage, he also contextualizes his analyses of social power and spiritual powers with serious statements. For example, he writes, "I am not speaking hastily or lightly here but as one who has passed most of his life in confrontation with this question and this power."[12] Accordingly, Ellul's abbreviated answer to the question "whether technology is not also one of these powers" ought to be framed by the latter passage's seriousness.[13] It ought to urge readers to search for other places in Ellul's dialectical writings where he is more forthcoming about his understanding of power(s) and technology because there are such places. As he writes in *Théologie et Technique*, "The spirit without its conditions of existence, without its incarnation, and since we speak of Technique, without means, is nothing." He concludes, "It appears to [me] that [the] spirits which are at the heart of the technological movement are the spirit of power and the spirit of lies."[14] The following sections will expose and name these two vicious powers at play in Ellul's dialectical writings on technologies.

THE SPIRIT OF POWER, MIGHT, AND CONQUEST

When The Spirit of Power appears in Ellul's writing, technology often appears, but the relationship between this power and technologies is not always explicitly articulated. Decoding Ellul's various names for The Spirit of Power shows how The Spirit of Power, technology, and individuals/institutions interact in Ellul's dialectical writings on technology. The Spirit of Power emerges in Ellul's writings from a mixture of sources inspired by social phenomena, literature, spoken and written commonplaces, other theologians, Ellul's own theology, and scripture.[15] It requires extensive reading to see how present The Spirit of Power is in Ellul's writing. He tends to use diverse names in his sociological and theological writings to describe the one spirit. For example, in *Propaganda*, Ellul often refers to The Spirit of Power as "the drive for conquest," writing, "To the extent that members of a group believe their ideology to represent the truth, they almost always assume an aggressive posture and try to impose that ideology elsewhere. In such cases ideology becomes bent on conquest. The drive toward conquest may arise within a society as a conflict between groups [. . .] or it can aim at targets outside, as a nationalist ideology will."[16] First, it is possible, though not verifiable through references in Virilio's work, that passages like these influence Virilio's work on the human drives for speed and power that have been examined so far. Either way, Ellul and Virilio move their understandings of the postlapsarian drive for conquest, speed, and power, in different, though complementary, directions. In *To Will and to Do*, for example, Ellul writes about the will to power: "Adam's break with the Father brings about negation, chaos; and everything is in danger of returning to nothing—whenever this freedom is no longer in love, it is a great confusion. Once love has disappeared through the will to power, the significance of everything changes."[17] In other texts, Ellul mixes these terms together with allusions to early Christian thought. *Violence*, for example, combines Ellul's history, sociology, theology, and ethics to name The Spirit of Power in connection to the Fall and technologies:

> It is the very old attitude, described at length in the Bible, of human pride, which has always tried to break away from God; it is the "will to power." What is really new is that, instead of realizing that all this is the manifestation of an ancient evil (*potestatis cupido, avitum malum!*) [*lust for power, ancestral fault!*] and of the lost condition of man, who seeks a way out of his complex of "despairing power"—what is really new is that Christians today find this prideful attitude excellent and consonant with the dignity of man. By breaking loose from God's ancient tutelage, they believe, man has graduated into an authentic situation [. . .]. There is only one explanation of this "conversion" of Christians to the spirit of power; namely, conformism to today's technological society and awe at the advances of science. Obviously, such conformism entails a series of results. In the first place, it entails rejection of the ethical consequences of man's condition as a sinner in need of forgiveness.[18]

Here, in one passage, Ellul knots together various theologies, concepts, and motifs to expose a connection between The Spirit of Power and technology in the modern world, both in relation to and beyond Christians and churches. To mark other places where Ellul similarly ties together theologies, concepts, and images to name The Spirit of Power, he often describes it in a style similar to the aforementioned passages, but uses names and characterizations, which differ depending on the circumstance. For example, Ellul often characterizes The Spirit of Power as "covetousness and spirit of conquest"; "the power which is against God, covetousness-impotence, the spirit of might and conquest"; "the spirit of power and conquest, [the] spirit of independence and surfeit"; "lust for power"; "the spirit of power, [the] spirit of self-interest, [the] spirit of the will to dominate"; "Dionysian man"; "an unleashed, torrential, Dionysian power, a superpower, covering everything, making use of everything without order or restraint, devastating and depleting the earth"; "Dionysian explosion of technical power"; "the madness of power or hubris"; "the attitude [of Greece and Rome]"; "the possessive Eros"; and so on.[19] Considering these various iterations of The Spirit of Power in Ellul's writing, there are two possible reasons for this variation in names, terms, and imagery. The first is that it could be a lack of precision in Ellul's writing. The second and more likely reason is Ellul develops various renderings of the concept so as to communicate its constant but difficult to name presence in the world. In other words, Ellul's varied cognates and images for The Spirit of Power function in a way that is similar to his understanding and use of sociological *myth* and theological *myth.* The Spirit of Power moves through his sociology, theology, and ethics of technology as a mysterious, but formidable force; a cryptic phrase, name, and word developed to communicate a truth, which cannot be communicated literally or directly. In this way, Ellul's exposure and naming of The Spirit of Power in his dialectical writings on technologies is a part of his demythologization/remythologization effort. Whenever the Spirit of Power and its iterations appear in Ellul's dialectical writings, *technique*, *technology*, and *propaganda* also tend to appear.[20] In *Ethics of Freedom*, for example, Ellul writes, and it is worth quoting him at length here:

> The moment that freedom is based on the desire for conquest, and prayer, asking, and waiting seem to be servile attitudes, at that moment freedom is the same delusion as the choice of good or evil by fallen Adam. [Once] Adam is separated from God even his determination of good is evil. In the same way, the freedom which man has achieved by conquest is intrinsically bondage. We are not talking lightly here. We are not playing with words. We are not engaging in philosophical discussion. Revelation shows it to us and history confirms it. In human history there has never been a conquest of freedom without destruction, blood, and the enslaving of others. Always, too, the result has immediately been bondage for those who thought they were achieving liberty. The adventure of modern technology teaches the same lesson. The bondage does not come merely from disregard for the others who are obstacles to our desire for conquest. It comes also from determination by the object [. . .]. The trigger of my desire for conquest, the object itself, becomes the meaning of my life, the catalyst of my intelligence, the

> nerve of my energies, the pole of my orientation. In other words, I cease to have any other life-possibilities. The desire for conquest puts me on a strict course that I no longer have the power to abandon. It steers me towards the object. It takes the fun out of my life and crushes me with implacable seriousness.[21]

Ellul concludes: "One might say that the very essence of the desire for conquest is the multiplication of means. Now in our view the Bible teaches that the increase of means is destructive of freedom" even if "the man who justifies himself endlessly can persuade himself that the mediated relation he has with others is love."[22] Compare this to *Presence in the Modern World*, where Ellul writes, "When we say that technique seems neutral what we mean is that deep within ourselves we all think it is good. Technique is good because it extends the power of human beings, and means are justified today by the power that they give to human beings. This is the meaning of the self-justification."[23] Certainly, The Spirit of Power is moving in Ellul's early to late dialectical writings on *technique*, *technology*, and *propaganda.*

By the time *Hope in Time of Abandonment* is published, Ellul becomes noticeably direct about his view of social power and spiritual powers. "As far as power is concerned," he writes, "Exactly nothing has changed since Genesis. When we take into consideration the actual facts of human power, how could we be so absurd as to turn the biblical myths upside down? Man's power is always a consequence of the break with God. It is always the deed of Cain, of Babel, of Nimrod, and of Mammon. There is no other."[24] Then, quite tragically, he laments,

> I know that in writing this I am going counter to the contented purring which justifies the 'greatness' of modern man, but what can I do? Torture, exploitation, dictatorship, and the universal destruction of nature constitute the essence of the power of man. The blind sycophants ignore fact. They ignore the concrete reality of the facts as well as the internal structure of the phenomenon.[25]

With all of these passages in view, it is evident that power(s), especially The Spirit of Power, animates Ellul's early to late writings on technologies. Yet, the precise way in which this spirit functions in relationship to technologies—Ellul mentions "the internal structure of the phenomenon" in the last quote—constitutes a complex process, which requires a detailed look to see how it works. The next step, then, is to look at the presence of The Spirit of Lies in Ellul's work on power, powers, and technology.

THE LYING SPIRIT

Whereas The Spirit of Power animates *technique*, The Spirit of Lies animates *technology*, and this mutually reinforcing relationship produces *propaganda.*

Whenever the discourse of *technology* justifies the rational process of *technique*, and, with the cruel aid of *propaganda*, *technique* becomes the one best way to think, act, or make efficiently, The Spirit of Power and The Spirit of Lies have used and damaged individuals and institutions that have idolized technology.

The Spirit of Lies, which can also translate *The Spirit of Falsehood* or *The Lying Spirit*, emerges in the early- to mid-1970s as Ellul started writing *Technological System* and *Théologie et Technique*. Considering chronological development and textual connections it could be argued that Ellul's understanding of The Spirit of Lies became a theological basis for *Technological Bluff*. Ellul opens the section describing The Spirit of Lies in *Théologie et Technique*, using the word *bluff* in the following way:

> The second spirit which is at the heart of Technique is the spirit of lies. This may seem surprising and unacceptable. Is not Technique derived from science, and does science not have as its objective the truth, and does it not indeed supply certain truths? Or is Technique not a practice, and is it not a practice that is the only acceptable test of truth, obliging ourselves to strip ourselves of a deceptive idealism, confronting us with reality? Here one cannot cheat, one is put at the foot of the wall. Technique is honesty itself, that of performance. Today we can make this speed, this construction, this operation, there is no bluff possible—it is yes or no—what could be more honest! And yet we shall see that the other instigator of Technique is the spirit of lies.[26]

This passage theologically mirrors in style and content the introduction to *Technological Bluff*, Ellul's last sociological work on technology:

> I have entitled this book *The Technological Bluff*. Most who see the title will react sternly. Technique is an area in which bluff is not permitted. Things are plain in technique—either one can or one cannot. When it is said that people will walk on the moon, shortly afterward they do. When it is said that an artificial heart can be inserted, it is done and it works. Where is the bluff? But the problem is one of language. American usage has implanted in our minds the idea that the word *technology* refers to actual processes. This is the way the media uses the term. But in a strict sense technology is discourse on technique. It involves the study of a technique, a philosophy or sociology of technique, instruction in technique [. . .]. I am talking about technological bluff, about the gigantic bluff in which discourse on techniques envelops us, making us believe anything, and far worse, changing our whole attitude to techniques.[27]

Such textual connections are important to highlight not only for clearer readings of how theology informs Ellul's sociological writings on technology, but also because they push at general readings of Ellul's dialectical work, which keeps the sociological-historical and theological-ethical texts too separate. Here, as elsewhere, there are strong indications in Ellul's work, on both sides of the dialectic, that Ellul's sociological-historical and theological-texts are far more connected, and informed by each other, than many Ellul scholars grant. In fact, *Théologie et Technique* offers further theological frames for *Technological Society, System*, and *Bluff* when Ellul characterizes the distinctive powers of The Spirit of Power and The Spirit of Lies in the following way. It is worth quoting Ellul at length here given how the passage theologically frames these three sociological texts, and substantiates the readings of Ellul's dialectical writings on technologies developed in this and other chapters:

> As [it is] for the spirit of power [. . .] this spirit of [lies] exists by the conjunction, the bond without break, the entanglement between the spirit, the will of man, and then the development of means as a system. Not one without the other. Man, with his will for self-justification, duplicity, the diversity of his sincerities, his structures [of] contradictory strata, his false consciousness [. . .] etc., a lie which is in no way a matter of morality, but which seems linked to the very structure of [man's] psyche, and then, the other factor, Technique developed as a system, producing a rigorously illusory and deceitful [or trick-playing] universe. I could say that this permanent process of developing Technique consists in its being taken for what it is not [. . .]. It is man who receives this representation and it is he, the accomplice, who develops the luminous superstructure of the system, all too happy to participate in this lie, because he cannot tolerate an impeachment of the work of his hands. It is this relationship that is the spirit of lies.[28]

Ellul immediately clarifies "lies" is not a synonym of "negativity" nor "the spirit of negativity" here. Rather, The Lying Spirit's "lies" ought to be thought of as deceptive plays on truth, a "doubly positive" discourse.[29] In other words, The Spirit of Lies, through *technology* and *propaganda*, activates *myth* to institute the sacred order of *technique*, using dangerous riffs on the truth, discursive tricks, and doubly positive affirmations to justify the sacralization of *technique*. It gives the *Lie* in *The Spirit of Lies* a spatiotemporally justifying connotation in that *lies* or *lying* means either *falsehood* or *positioned.* As far as self-justification is concerned, *lies* and *lying* mean both, especially in relation to *technology* and *propaganda*'s idolization of *technique*.

The rough-draft nature of Ellul's writing in these passages produces some ambiguity insofar as Ellul tends to use "Technique" and "technical" to characterize what he typically would describe with the terms *technology* or *propaganda*. With that said, the discursive qualities of The Spirit of Lies as Ellul describes it throughout these passages imply that his use of a capitalized

"Technique" connotes the unified phenomenon of *technique*, *technology*, and *propaganda*. In fact, Ellul continues, writing with sarcasm, "Technique is innovative" and "it always brings the positive: products, possibilities, solutions, everything in it is possibility." But he also replies with terse sincerity, "This is the first lie: for [Technique] destroys as much as it produces." In this lie, "The enormous technical negative is completely obscured, veiled, so that no one realizes it." At that moment, "The spirit of lies appears as soon as Technique is evaluated in terms of its positivity."[30] In this way The Spirit of Lies animates *technology* to justify the rational process of *technique* and The Spirit of Power. This produces *propaganda* and makes *technique* "the norm of norms" in a society. When this happens, *technique* becomes "a *manifest* sacred" because "manifestation is indeed the essence of Technique, it simply does not exist at all without manifestation, it is made only of manifestation. And it is here that we detect in this domain the spirit of lies."[31] It "plays exactly the role and the function of the sacred, but it is not [sacred], it is only a parody, a counterfeit, an artifice of the sacred."[32] Rather roughly, these passages coalesce to demonstrate the following: *technology* is animated by The Spirit of Lies to institute the sacred order of *technique* as described in *New Demons*, and this self-justifying dialectic manifests—it makes (*tékhnē*)—good and evil in the world, as described in *To Will and to Do*.[33]

Precisely in relation to *myth*, the sacred, and The Spirit of Lies, Ellul elaborates, writing, "Technique is provocative of the *myths* of the future. But it thus reveals its spirit of lies, for it is not the future *of man* that is evoked."[34] Rather, "The only future we are talking about, is the future of Technique itself, it is its own development." In "the rigorous conviction of all the technicians—and who, in our Western world, is not a technician?—we see the circularity of the lies of Technique."[35] So much, in fact, "We see Technique draping itself in soteriology!"[36] By parodying true justification and the true soteriology to which the truth refers, The Spirit of Lies manufactures a false soteriology. This substantiates the thesis of the previous chapter, in which Ellul also writes on the subject of justification, soteriology, *technique*, *technology*, and *propaganda*: "It is [Technique] that places us before this option: 'Me, more—or nothing.' So there is no other way, no other history, no other possibility. Here is excluded the statement of Jesus: 'I am the way.' "[37] To reiterate what was suggested in the previous chapter here, then: when Ellul argues "the one best way" of *technique*, *technology*, and *propaganda* is the "formula to which [modern] technique corresponds," he is describing the dangerous sociological obverse of a theology of salvation by faith alone in Jesus Christ, who, for Ellul, is the *only way*, truth, and life.[38] The Spirit of Lies, *technology*, and *propaganda* animate this parodic and dangerous one best way, and as a consequence, *technique* and The Spirit of Power retain their destructive power in the world. Meanwhile, individuals and institutions remain immersed in the illusion that *they* are in control of their idols, and terrible, destructive, power is produced.

The aforementioned descriptions of The Spirit of Lies are theoretical and they do not offer much insight into the biblical, theological, or literary sources Ellul uses for his exposure and naming of The Spirit of Lies. However, he does conclude the section on The Spirit of Lies in *Théologie et Technique* in the following way: "Technique manifests the spirit of power and the spirit of lies, it is its most complete form. It is entirely inspired by it."[39] However, "I do not formulate this as a judgment against Technique: it is absolutely fundamental to know what is at the heart. Now, by the conjunction of a biblical reading, a theological reading, and a sociological reading, this is what is significant."[40] He alludes to biblical, theological, and sociological sources for The Spirit of Lies, then, but does not provide details about these sources in *Théologie et Technique* as he does for The Spirit of Power. Nevertheless, these concluding comments, as well as the earlier-given passages, do show how tightly woven together the biblical, theological, and sociological elements are in Ellul's dialectical writings on technology.

As far as biblical sources are concerned, the likely ones for Ellul's naming The Spirit of Lies are 1 Kings 22:19–23 and 2 Chronicles 18:18–21, in which "a lying spirit" speaks through the mouths of prophets who speak deceitfully positive words to the king of Israel. In contrast, the prophet Micaiah is the only one who will prophesy the truth as it is given to him by God even if it displeases the king. Eventually Micaiah is imprisoned for his honesty and the king and false prophets are abandoned to their folly. Given Ellul wrote a commentary on Second Kings in *The Politics of God and the Politics of Man*, and given his ironic commentary on the folly of royal power in *Reason for Being*, it is probable, though it cannot be said for certain, that these scriptures were used in Ellul's naming The Spirit of Lies.[41] The incomplete nature of *Théologie et Technique* implies more than it does evidence directly some of the sources and textual connections examined in this section. It also should be noted that Ellul's understanding of the principalities and powers changes over time.[42] But throughout Ellul's dialectical writings on technologies, and especially in *Théologie et Technique*, The Spirit of Lies is taken seriously for the parodic power it has in the fallen world, especially in relation to *technology* and *propaganda*. The Spirit of Lies is not the Devil (cf. John 8:44) as such, but rather, like The Spirit of Power, it is a mysterious spiritual power at work in the unified phenomenon of *technique*, *technology*, and *propaganda*, which people use, and are used by, because they lust after, worship, and idolize power.

All of the aforementioned contexts constitute a complex process, which can be described as follows. The Spirit of Lies animates *technology* and justifies The Spirit of Power's animation of *technique*; this mutually powering interaction produces *propaganda*, activating sociological *myth* to institute the sacred order of technology; and, individuals and institutions enter into this process to manifest and amplify these spiritual powers for the social power it seems to promise. But, and here is the truly vicious parody of the power process: by using these means for the end of social power, the individuals and institutions become the means for these spiritual powers in the end. Then, the individual and institution are used, duped, and stripped. "Here we find an important revelation concerning the city," Ellul writes.[43] "The work of man which was meant to protect and cover him, to give expression to his desire for power, is in reality naked and miserable. The city is naked, without ornament and beauty—like a dead body abandoned in the fields. Despite all of man's best efforts, she needs other clothing, other ornaments; the work is like its maker."[44] The "worker's folly is to believe that the work of his hands could be helpful and effective in his adventure, could protect him, whereas the work itself needs to be clothed."[45] The parodies of The Spirit of Lies are not a joke for Ellul. They are serious threats. Idolatry, though most tend to underestimate its power and appeal and treat it like a vague concept, a vice that can be attributed only to other people, and thus something to be laughed at when mentioned, is powerful and appealing. And, as Ellul warns, giving into that power and appeal will cause spiritual and physical death.[46]

When works become the externalization of a fallen person's desire for power, and as a consequence that person joins with The Spirit of Power and The Spirit of Lies to see those works increase in power, a destructive alienation from the self occurs, and keeps occurring, until the destruction reaches an end: either a metaphorical or literal dead body, abandoned, in a field. In *Ethics of Freedom*, Ellul writes, "Now the freedom that Christ gives is first of all a liberation from the flesh (Romans 6:6)," but an "invasion of the flesh means submission to what has become an alien power even though it is ours."[47] In the latter case, "Man [becomes] occupied by the flesh, by covetousness-impotence."[48] For Ellul, the question of discernment of the spirits is framed by questions about the clothing of the person, The Spirit whom indwells, or the spirits which possess, a person. He writes, "Already we can live according to the Spirit who is given to us, for the Spirit now dwells in us, not the flesh. Our body is the temple of the Holy Spirit. We are possessed, not by the flesh, but by the grace of God."[49] The next chapter will show how Ellul's theology and ethics highlight a critical, theologically and ethically framed, question: with whom or what will we be filled, with whom or what will we be clothed, and with whom or what will our works be adorned? Here, Ellul offers an answer in *Théologie et Technique*, which prefigures the next chapter. He writes, "You have to find the root and act on that level. The concrete consequences will come of themselves when the central point has been modified. It is useless to take plums, to remove the acidic fruits of a wild plum, and to suspend the plum on the plum tree in question." That "simply leads to the decay of plumbs."[50] Instead, "The only correct attitude is to graft the wild plum, but it will bear fruit only one year later."[51] The fruits of The Spirit take time to grow, but they bear good fruit, blooming into a multiplicity of fruitful thought, action, and making. The power and parody of The Spirit of Power and The Spirit of Lies appear to guarantee, immediately, efficiently, the fulfillment of our fallen desires. What will we choose: the fruitful efficacy of the Holy Spirit or the seemingly efficient means of The Spirit of Power and The Spirit of Lies? I take Paul's warning in Romans 11:24 as an essential scriptural frame for our answer to this question: "For if you have been cut from what is by nature a wild olive tree and grafted, contrary to nature, into a cultivated olive tree, how much more will these natural branches be grafted back into their own olive tree." Neither the cultivated nor the wild branch or plum can consider itself invulnerable to the temptations of power, parody, and idolatry.[52] As external forces press upon us from the outside, our internal abysses are calling out to be filled. With what or whom will we be filled?[53]

THE ACCIDENT'S SUBSTANCE

Ellul's writings on technology are best understood if his account of The Spirit of Power and The Spirit of Lies is appreciated. It formulates and helps to answer an important question for theology and ethics today: How can devastating social power and spiritual powers be exposed, named, and characterized in ways which resist the destruction and deception their idolization causes in the world? The reading developed in this chapter clarifies the purposes of Ellul's having written dialectically on technology.

Regarding Ellul's ethics, a Christian ethic only can be developed after the abyss of creaturely fallenness, with its self-justifying lusts for power and coveting of protection and provision in separation from God, has been peered into. Note especially how the themes of nakedness and clothing have emerged in this chapter because the next chapter examines the importance of these themes for Ellul's dialectical writings on technologies. The conclusion of this chapter, however, demonstrates why, for Virilio, the abyss of separation from God deepens with accidents. Typically, accidents are viewed as events or phenomena that appear by chance and cannot be avoided—car accidents, for example. But Virilio's dromological phenomenology offers a different view of accidents. For Virilio, accidents expose things that are always present but often hidden behind the facades of materiality: substantial and precarious power(s), which ought to be identified, and in some cases, resisted or avoided.

Other than being known as a philosopher of speed, Virilio is most known for his cryptic axiom *the invention of the ship is the invention of the shipwreck.*[54] He developed the axiom through engagement with Aristotle's studies of the accidental and substantial qualities of a person or a thing, Aristotle's attendant observation that there is no science of the accident, and Nicholas of Cusa's theological reframing of these Aristotelian motifs.[55] Virilio responds to Aristotle and Cusanus by demonstrating why there should, in fact, be a science of the accident in a modern world overrun with technologies, which frequently instantiate accidents.[56] Via phenomenology, Virilio shows how technologies, and the spaces shaped by technologies, both of which most people perceive as commonplace and unthreatening, actually tend to be pervaded by a vicious militancy, and at times, can manifest either the illusion or actual presence of ghosts, phantasms, spirits, or demons. For instance, in *War and Cinema*, he writes the following, and it is worth quoting him at length, given how it draws together many of the critical themes, which have been examined and commented on thus far:

> The First World War brought an end to the privileged relations between old religions and young military-industrial states. These states, founded as in the Soviet Union upon overt violence, needed to create a new consensus in order to be accepted by the greatest possible number (in order, that is, to become legal)—hence the urgent necessity of imposing replacement cults on the masses. The mystical, scientistic, materialism of the nineteenth century had, in assuming effective power, become a question of working "miracles" of science by means of technology. Paradoxically, the supposed advent of Reason in History became a cultural farrago, a technophile syncretism, complete with a demonology and inquisitorial apparatus of which the personality cult is one of the best-known aspects. By the middle of the nineteenth century, recent scientific discoveries and technological applications had unexpectedly brought mesmerism and Swedenborgism back into fashion into Britain and the United States. It was the hour of the new illuminati, for whom "the spirit" (God, the Eternal) was a magnetic fluid, and electrical phenomenon, kinetic heat. In 1845 the first spiritualist rappings were heard in the home of the Fox sisters in Hydesville. In the United States séances, dictations from beyond the grave, and others forms of hypnotic or medium-induced communication were used by individuals and religious sects "to move back in time or across space without too much effort," as Polytechnic director Colonel de Rochas put it in 1900. The American Catholic Church itself did not completely disapprove of such goings-on. For it was a question of seeking out *new vectors of the Beyond*, after the provisional decline of great religions and their loss of prestige to the state.[57]

Note especially in the middle of this passage where Virilio communicates in one compact sentence ("It was the hour of the new illuminati") how the material cinema environment (*phusis*) harnesses a kinematic energy (*kīnēsis*), which illuminates (*phōs*) a dark room filled with people and magnetically draws in (*drómos*) the audience to worship a new, dangerous, perhaps demonic, kind of spirit.[58] These phenomenological descriptions explain why he describes the cinema as representative of a "flourishing 'ghost industry'" and assembly point for "black masses."[59] These are negative images of believers gathered for mass in cathedrals filled with light. If the human drive for speed and power tends to dissolves creation and creature's mass, then the Mass held and masses gathered in the actual cathedrals for worship can stand against such dissolution.[60]

Virilio's analyses of accidents, or what at certain points he calls "miracles in reverse," "secular miracle[s]," "inverted miracle[s]," "revelation[s]," and so on, help him tear off the material masks of technologies to expose the violent power(s) lurking behind the veil.[61] As such, he does not need to provide a systematic account of his theology of the principalities and powers as Ellul does. Connecting the basic social phenomena of cinema and film to war, he will observe, "There is no war, then, without representation, no sophisticated weaponry without psychological mystification. Weapons are tools not just of destruction but also of perception—that is to say, stimulants that make themselves felt through chemical, neurological processes in the sense organs and the central nervous system, affecting human reactions and even the perceptual identification and differentiation of objects."[62] Or, more simply, "For [people] at war, the function of the weapon is the function of the eye."[63] The faith (i.e., idolatry) element comes into play here when Virilio points out that, "Apropos the visibility now at issue in the religious as well as the political order, let's not forget that the sight of the path, the line of sight used [when aiming a rifle or any other weapon], that allows man to attain his target, used to go by the name, in French at least, of 'line of faith' [*ligne de foi*], though the term is today censured and replaced by the term 'line of aim.'"[64] In the way that we aim with our guns and our cameras, we show that each aim, pull of the trigger, or push of a button is, in fact, an act of faith. In some cases, this act (*poiēsis*) of faith has significant consequences, in others insignificant. Whatever the consequences, however, the faith we put in our sight, and where we tend to aim our gaze, is a good indication of what we do and do not worship and the effects our worship will have. With menacing sentences, theories, and applications such as these, Virilio is able to show convincingly why it is, that, however one thinks about technologies used within this or that environment—in this case whether it is a battlefield or a cinema—once technologies start increasing speed (*drómos*), harnessing motion (*kīnēsis*), manipulating light (*phōs*), and driving people away from God and God's creation (*phusis*), the more accidents will appear, and the less it will be possible for people to sense a clear difference between thought, action, and making within the defined categories of everyday life, violence, war, or cultic celebrations of ghosts, phantasms, and demons. In war, as in cinema, as in so many other commonplace technologies, the stable and fixed forms of the physical environment blur together in a collision of light, motion, speed, violence, and phantasm.

It is not just for rhetorical affect, then, that Virilio points out in *War and Cinema* that some kinds of film stock commonly used in the twentieth-century were actually flammable. As he writes, "Just as the nitrocellulose that went into film stock was also used for the production of explosives, so the artilleryman's motto was the same as the cameraman's: lighting reveals everything. By 1 October 1914 anti-aircraft artillery was already combining guns with searchlights."[65] Virilio really thinks, and is able to show, how something as common as film stock can be just as explosive of artillery.[66] He skillfully unveils the ghastly and ghostly substance(s) of peoples' uses of technologies, which most people would rather not perceive and many would choose to deny. But it is through such exposure that he can show so clearly how and why what most people tend to perceive as *just accidents* actually can reveal the disturbing and dangerous substance(s) of a person, thing, or environment, which taints all thought, action, and making in the world. "The information bomb" is one of Virilio's most well-known examples of this common theme in his work. For example, he writes: "News is dynamite, information explodes like a bomb, opinion polls or war propaganda are time bombs."[67] The "administration of fear" is another: "This information managing (generating) capability that is gearing up to invade the imaginary of populations held in thrall by the proliferation of screens [perfectly] typifies the globalisation of 'affects'—this sudden synchronisation of collective emotions greatly [favors] the *administration of fear*."[68] Yet, whatever the example used, Virilio describes himself as a "revelationary," and he frequently reveals, through his study of the accident and accidents, or miracles in reverse, how people who use technologies always are closer to violence, war, the demonic, sin, and death than they would prefer to sense or admit.[69] This is why in one of Virilio's later interviews, he states openly that the word *accident* is a cognate for "sin."[70] What, then, one might ask, is the *sin* of the ship? The *shipwreck*, Virilio answers, and he frequently makes his case phenomenologically with his analyses of the ship and other technologies.[71] The point, for Virilio, is not that shipwrecks *always* happen because buildings ships, or ships themselves, are categorically sinful. His point is that when shipwrecks *do* happen, it reveals something of substance about the ship, and the building of the ship, that was excluded, either intentionally or unintentionally, from peoples' perceptions and imagination. The true sin is to witness thousands of shipwrecks and to ignore what those shipwrecks are revealing. The same goes for *any* technology or any accident.

The above produces three critical questions for ethics. First, at what point does an *accident* stop being called an *accident*? Second, at what point does a person, or people, become responsible for an *accident* that ought to have been foreseen? Third, can it be said that people who consciously ignore *accidents* ought to be considered *engineers* of *accidents*? For one of many examples, in *University of Disaster*, Virilio poses these questions in the following way:

> Even if it seems at least a little premature to apply the term "ENGINEERING" to natural disasters, this does have the merit of opening up the question of a practical collective intelligence of shared interests in the face of mega risks; and it does so at the precise moment when the catastrophe of a fatal convergence of all risks is on the agenda for BIG SCIENCE hijacked by the accredited experts of Progress at all costs, whose grab for power has wiped out all judgment, all rationality, turning research and development into a culpable passion more than a capable science; a sort of postmodern expressionism destined for the theatre of an epic science that would replay without letup the tragedy of the Garden of Eden, at the foot of the Tree of Knowledge—this time at the risk of expelling shameful Humanity, not from our "original paradise" anymore, but from Mother Earth.[72]

Grabs for power that ignore the revelatory *accidents* surrounding us dangerously replay the tragedy of the original grab for power, which took place at the Tree of Knowledge in The Garden of Eden. Adam and Eve ought to have taken responsibility for their power grabs at the Tree and repented. But they deflected the responsibility for their sins to their surroundings, blaming God, each other, and the serpent. In the same way, we ought to take responsibility for and repent of our power grabs, never deflecting responsibility and blaming God, other people, and our external environment. For this reason, combining Virilio's reading of Cain and Abel, the Fall, and how both of these readings theologically frame his understanding of the problems with and ethics for technologies, Virilio's phenomenological exposure of connections between power, powers, and technologies reinforces Ellul's. It focuses Ellul's dialectical and structural exposure of social power and naming of spiritual powers onto the particular technologies, which function in these dialectics and structures. Consequently, when Virilio says "speed is power," he is able to show how there are a plethora of powers hiding behind the technologies and accidents revealed by the power of speed.[73] For precisely this reason, his studies of how technologies can alter peoples' senses, the physics of creation (*phusis*), increase speed (*drómos*), harness motion (*kīnēsis*), and manipulate light (*phōs*), all become means by which Virilio can rip off the masks of technologies and expose the powers hiding behind them. Despite their different methods, Virilio, like Ellul, does not expose power(s) to categorically condemn all technologies. He exposes power(s) to show how they can be, or are being used, to legitimize and enhance forms of life that are destructive.

By this point, there have been various indications as to why justification or mediation are essential theological frames for how Ellul and Virilio perceive the problems with and ethics of technology. Now, with five of six chapters in view, these indications can be examined fully and the following thesis can be provided for the next chapter. First, Ellul emphasizes speech and justification and questions what people say about technology in his writing because he wants to create time and space for people's immediate hearing of God's Word. Second, Virilio emphasizes senses and mediation and shows people what they would rather not perceive about technology in his writing because he wants to create time and space for Jesus Christ to mediate peoples' senses of God, other people, and creation. Once this two-part thesis has been substantiated, I will draw out their most helpful, dynamic, and enduring ethics for thought (*epistēmē*), action (*poiēsis*), and making (*tékhnē*) in the world, which flow from their theologies of justification and mediation. The concluding sections will also examine the importance of both scholars' work for today, add to, and at certain points move beyond, the ethics derived from the writings of Ellul and Virilio.

NOTES

1. Ellul, *Theological Foundation of Law*, 107.
2. Ellul, "Chronicles of the Problems of Civilization: 1. By Way of a Brief Preface," *Sources and Trajectories*, 19–20.
3. Ellul, "Author's Preface," *Presence in the Modern World*, xv–xx.
4. For example, Ellul, *Prayer and Modern Man*, 151–152.
5. In fact, in Ellul, *Théologie et Technique*, 272, in a section that outlines the discernment of the spirits, Ellul mentions Karl Barth's discernment of Nazism's evil spirit.
6. Ellul, *Presence in the Modern World*, 93.
7. Dawn, *Concept of "The Principalities and Powers" in Jacques Ellul*, 170–192 explores how power(s) factor into Ellul's writings on technology but she only shows provisionally where they subtly appear in *Technological Society*, *Technological System*, and *Technological Bluff*. She does not examine all of Ellul's writings for more direct discussions of the relationship between power(s) and technology as this chapter will. Ellul provides a partial inventory of the powers in Ellul, "Dominions and Powers," *Subversion of Christianity*, 174–190. Right at the beginning, he mentions in

a footnote, "I have dealt with these two points at length" in *Presence in The Modern World* and *Money and Power*. Both are valuable sources for Ellul's early descriptions of the principalities and powers and how they animate social phenomena, including technologies and money. See, for example, Ellul, *Presence in the Modern World*, 95, where he writes, "A revolution is needed in a world in which [revolution] has become impossible, a revolution that attacks the deep structures of a civilization in which all efforts converge toward this one goal: to transform all human beings into swine, who for this reason can no longer receive the pearls of scripture. We must rediscover the meaning of human activity, the situation of means and ends, and their true place in the world entirely given over to the spirit of power, to dissolution, to pride of limitless means that absorb us without possible reserve."

8. Ellul, *Ethics of Freedom*, 153–154. The similarity between Ellul's descriptions of the powers and principalities in *Ethics of Freedom* and Barth's "Lordless Powers" in *CD I/4*, *Christian Life*, 213–232 is remarkable.

9. Ibid.

10. Ibid.

11. Scholarly debate surrounds the correct publishing dates for the respective English and French editions of *Ethics of Freedom*, but regardless of the correct dates, the content examined in this and other chapters is not effected. Ellul also mentions on page 151 of *Ethics of Freedom* that "there will be a fuller discussion" of his theology of the powers in his *Ethics of Holiness*, which is yet to be published or translated. The manuscript for this book currently is in possession of Ellul's son and presumably will be published at some point. This does not alter Ellul's writings on power and the powers as examined in this and others chapters because there already are available a good number of Ellul's writings on the subject. However, *Ethics of Holiness* could nuance or expand Ellul's account of the powers.

12. Ellul, *Ethics of Freedom*, 153.

13. Ibid., 152–153.

14. Ellul, *Théologie et Technique*, 273–274.

15. My dissertation includes extended commentary on the social phenomena, literature, spoken and written commonplaces, other theologians, Ellul's own theology, and the scriptures at play in Ellul's concept of the Spirit of Power. I cannot go as deep into those facets of his work here, however, given that it would be too much of a rabbit trail for the discussion at hand. But, for those who are interested in this material, see Morelli, *Either Technological Pessimism or Proleptic Christian Resistance: A Constructive Reading of Jacques Ellul's Dialectical Writings on Technology with Paul Virilio as an Evocative Interlocutor*, https://ethos.bl.uk/OrderDetails.do?uin=uk.bl.ethos.787402. I consider this work useful insofar as it offers theology and ethics a unique and powerful way to identify, expose, and name *exousia*, the principalities and powers of past, present, and future eras.

16. Ellul, *Propaganda*, 194.

17. Ellul, *To Will and to Do*, 60. See 7–8 and 190 for further discussion about power, the Fall, knowing good and evil, and technology.

18. Jacques Ellul, *Violence*, trans. by Cecelia Gaul Kings (New York: Seabury Press, 1969), 40–41. The Latin phrase "lust for power, ancestral fault!" is a reference

to patristic writings on ancestral fault, which similarly appears in Ellul, *Théologie et Technique*, 233. For some of many detailed discussions of connections between the Fall, covetousness, lust, flesh, sight, nakedness, The Spirit of Power, *technique, technology, propaganda*, and science, see, Ellul, *Ethics of Freedom*, 133–144; Ellul, *Humiliation of the Word*, 69; 97–98; and 206–207; Ellul, *Violence*, 162–155; Ellul, *Théologie et Technique*, 213–238 and 278–281; and Ellul, *Reason for Being*, 150–158.

19. Ellul, *Humiliation of the Word*, 258; 67; 97–98; Ellul, *Hope in Time of Abandonment*, 196–197; Ellul, *Ethics of Freedom*, 135; 217–218; 242–243; 265–267; 269; Ellul, *New Demons*, 78, 142–143 and 226–227; Ellul, *Théologie et Technique*, 221–222 and 255–256; Ellul, *Betrayal of the West*, 169. In this case, the full quote is "technique contains within itself, indeed it is itself, the madness of power or hubris." Ellul, *Meaning of the City*, 16–17; 164; 168–169; Ellul, *Betrayal of the West*, 40–41, 71, 75.

20. Ellul, *Ethics of Freedom*, 157–160.

21. Ellul, *Ethics of Freedom*, 266–267. Ellul also situates The Spirit of Power in history, connecting it with the "drive to conquest" mentioned in relation to Lenin and Hitler already quoted in this section. He writes: "Napoleon was totally determined by his conquests. He could not stop even though he might have wanted to do so. No one was less free than Napoleon. The same might be said of a Hitler or a Lenin. This is how it is with the desire for conquest."

22. Ibid., 267.

23. Ellul, *Presence in the Modern World*, 47.

24. Ellul, *Hope in Time of Abandonment*, 212–213. This links up with Ellul's commentary on Cain's building in *Meaning of City*, 3: "The seed of all man's questings is to be found in Cain's life in the land of wandering, always searching for a place where his need for security might be satisfied." See chapter 4 for further references and discussion. Also, Ellul says the same in *Freedom, Love, and Power*, 87 as he does here: "Humanity now does on its own what God freely gave, but which has been taken over. In other words, humanity has become possessed by a spirit of conquest. Humanity now accomplishes in domination what God wanted to be done in love. Hence, humanity is driven by a will to power. After this, whatever people want to say about the history of science and *technique* leaves me indifferent; because what really matters is whether or not, at the very heart of all this, there is the spirit of conquest and the will to power." So, "For myself, this is the crux of the theological development. I could add all kinds of details and elaborations, but it would not change anything fundamental."

25. Ibid.

26. Ellul, *Théologie et Technique*, 284.

27. Ellul, *Technological Bluff*, xv–xvi. Ellul's emphasis.

28. Ellul, *Théologie et Technique*, 284–285.

29. Ibid., 285.

30. Ibid.

31. Ibid., 290–291.

32. Ibid.

33. Ibid., 59; 261; and 290. In fact, Ellul refers to *To Will and to Do* in his discussions of ethics (pages 59 and 261) and *New Demons* in his discussion of the sacred on (page 290).

34. Ibid., 294. Ellul's emphasis.

35. Ibid.

36. Ibid.

37. Ibid., 193.

38. Ellul, *Technological Society*, 79–80. John 14:6.

39. Ellul, *Théologie et Technique*, 297.

40. Ibid.

41. Ellul, *Reason For Being*, 85.

42. A comparison of Ellul, *Presence in the Modern World*, 6–7 and 13–15 to Ellul, *Subversion of Christianity*, 180–186 characterize some of these changes.

43. Ellul, *Meaning of the City*, 99.

44. Ibid.

45. Ibid.

46. This theologically frames the shocking conclusion to *Technological Bluff*, "Terrorism in the Velvet Glove of Technology," 384–412 in which Ellul writes, "I am not talking about technical terrorism, although I might have shown that the growth and popularization of technical means has made terrorist acts much easier [cf. Jacques Ellul, 'La democratisation du mal [The Democratisation of Evil],' *Sud-Ouest* (September 6, 1981), as well as, in *What I Believe*, 59–62]. My subject, [is] the discourse of technique, or technology in the strict sense, and my point is that this discourse is terrorist" in the "abstract sense of molding the unconscious with no possibility of resistance."

47. Ellul, *Ethics of Freedom*, 135.

48. Ibid.

49. Ibid.

50. Ellul, *Théologie et Technique*, 311–312.

51. Ibid.

52. Theologically and ethically speaking, the fuller scope of Romans 11:11–23 resists a reading that Jews or Gentiles are better equipped to be, become, or stay a part of the "cultivated olive tree." It all comes down to a question, as Paul says, of the person who stands "through faith" or falls with "unbelief."

53. Ellul, *Reason for Being*, 206–207; 211–212; and 300–303 elucidate this point in Ellul's terms.

54. Virilio, *The Vision Machine*, 28; Virilio, *Open Sky*, 40; Virilio, *Pure War*, 45; Virilio, *Politics of the Very Worst*, 89; Virilio, *Polar Inertia*, 54; Paul Virilio, *Landscape of Events*, 54; Virilio, *City of Panic*, 28; and Virilio, *The Original Accident*, 5 and 10.

55. Virilio, *The Vision Machine*, 28; Virilio, *Pure War*, 45; Virilio, *The Original Accident*, 5 and 10; Virilio, *Open Sky*, 17; and Paul Virilio, "The Third Interval: A Critical Transition," *Rethinking Technologies*, eds. Verna Andermatt Conley trans. Tom Conley (Minneapolis: University of Minnesota Press, 1993), 9.

56. Virilio does not cite Ellul in his analyses of accidents but Ellul also wrote about accidents and technologies. Their methods for writing about accidents differ,

however, in that Ellul's is, of course, sociological or existential, whereas Virilio's is philosophical or phenomenological. The text of Ellul's that seems to have influenced Virilio most in his writings on accidents is *Technological Bluff*, 9–10; 33–44; 46–47; 56–63; 68–85; 92–109; 113–121; 149–150; 160–169; 176–178; 186–187; 191–196; 210–211; 235–258; 265–266; 277–288; 194–300; 318–321; 375–376; 382–385; 396–399; and, 403–412. For other possible influences see Ellul, *Technological System*, 63; 280; and 288; Ellul, *Meaning of the City*, 153; Ellul, *Hope in Time of Abandonment*, 76–77 and 195–196; Ellul, *Ethics of Freedom*, 404–405 and 492; Ellul, *New Demons*, 204; Ellul, *Living Faith*, 3–7; 12; 217–255; 229–231; and 266–271; Ellul, *Reason For Being*, 62; 64; 96–9; 152–153; and 207–208.

57. Virilio, *War and Cinema*, 36–37. Virilio's emphasis. It is helpful to compare Virilio's phenomenological analyses of cinema and sight here to Ellul's more sociological analyses of festivals and sound in *New Demons*, "Secular Religions," 122–165. Ellul writes, "Music, drugs, incenses, violence, sexual freedom: festival. It is indeed noteworthy that our modern thinkers, revolutionaries, anti-technologists, renovators of mankind, preachers of liberty, all put their hopes in utopia, as we have seen, and in the 'festival': happenings, barricades, confrontations, sit-ins, fancy dress, theatre of participation, pop festivals, deafening and intoxicating music, audiences swaying with the rhythm of the drums, a participation which becomes identification. Each participant does himself in with sound, with fatigue, with drugs, with violence, with cries, with impressions received in common" (142).

58. See Virilio, *The Great Accelerator*, 89 for a similar discussion of "kinematic" energy.

59. Virilio, *Politics of the Very Worst*, 89; Virilio, *Polar Inertia*, 54; Virilio, *Crepuscular Dawn*, 148; Virilio, *City of Panic*, 29; Virilio, *Desert Screen*, 20 and 34; and Virilio, *The Accident of Art*, 63. Compare these passages to Ellul, *Living Faith*, 12: "The true miracle [supposedly] takes place without any help or intervention from God. Open-heart surgery is a miracle, and walking on the moon is a miracle, but there was no need for God to manifest his presence in either. The only domain he has left is that of chance, of unforeseen accidents. When an operation fails, there was an imponderable factor frustrating the expected result. In other words this God is nothing more than the God of obstacles, failures, and bad risks. The real expression should be [contra Saint Ambrose]: 'I dressed his wound, and God prevented him from getting better.' Are we dealing with an evil God, then? A jealous God who makes human power come to grief, the God of roadblocks and catastrophes? For the moment let's not go so far. But you can see that in the end that's what we'll be left with."

60. In fact, such negative imagery of the dissolution and fascistic bonding of mass and masses caused by mass transportation technology appears throughout Virilio's work. For examples, see Virilio, *Speed and Politics*, 31; Virilio, *Lost Dimension*, 87; Virilio, *Bunker Archaeology*, 11; Virilio, *The Art of the Motor*, 80; Virilio and Lotringer, *Pure War*, 45; Virilio, *Polar Intertia*, 83; Virilio, *The Information Bomb*, 140; Virilio and Armitage, *Virilio Live*, 138–139; Virilio, *Ground Zero*, 14; Virilio, *City of Panic*, 55 and 129; Virilio and Lotringer, *Accident of Art*, 83; Virilio, *Negative Horizon*, 117–120; Virilio, *University of Disaster*, 89–90; Virilio, *The Futurism*

of the Instant, 16; Virilio, *Administration of Fear*, 73–74; and, Virilio, *the Great Accelerator*, 31–35.

61. Virilio, *Politics of the Very Worst*, 89; Virilio, *Polar Inertia*, 54; Virilio, *Crepuscular Dawn*, 148; Virilio, *City of Panic*, 29; Virilio, *Desert Screen*, 20 and 34; and Virilio, *The Accident of Art*, 63.

62. Virilio, *War and Cinema*, 8.

63. Ibid., 26. Also, Virilio, *Desert Screen*, 34 and 39.

64. Virilio, *Art as Far as the Eye Can See*, 107.

65. Virilio, *War and Cinema*, 20.

66. Virilio, *The Vision Machine*, 70; Virilio, *Art as Far as the Eye Can See*, 107; Virilio, *Virilio Live*, 38 and 186.

67. Virilio, *The Art of the Motor*, 24. Also Virilio, *Information Bomb*, 63 and 108: Virilio and Lotringer, *Crepuscular Dawn*, 135–136 and 170; Virilio and Lotringer, *Accident of Art*, 84–85; and Virilio, *Original Accident*, 35–36.

68. Virilio, *The Original Accident*, 17. Virilio and Richard, *Administration of Fear*, 14–15 and 18–20. Virilio's descriptions of "the administration of fear" and the democracy of emotion are quite similar to Ellul's descriptions of the democratization of evil as mentioned earlier.

69. Virilio, *Pure War*, 223; Virilio, *Grey Ecology*, 43; Virilio, *Negative Horizon*, 61; Paul Virilio, *Popular Defense and Ecological Struggles*, trans. Mark Polizzotti (New York: Semiotexte, 1990), 29–30; and, Virilio, *Lost Dimension*, 182.

70. Virilio and Lotringer, *The Accident of Art*, 87. Also, Virilio, *Lost Dimension*, 96. Virilio, *Pure War*, 47 and 202; Virilio, *Virilio Live*, 106–107 and 154; Virilio, *The Original Accident*, 6–7; Virilio, *University of Disaster* 124 and 133–134; and Virilio, *Administration of Fear*, 71 and 81–82.

71. Virilio and Armitage, *Virilio Live*, 106–107. "Why do they always conceal *the original sins* of these techniques, whereas shipbuilding was furthered by making ships waterproof, and the aeronautic industry was furthered by making engines and the monitoring of air space more reliable. Why don't we have such people in the realm of digitization?" Emphasis added.

72. Virilio, *University of Disaster*, 123.

73. Virilio, *Politics of the Very Worst*, 15–16; Virilio, *Virilio Live*, 26; Virilio and Lotringer, *Crepuscular Dawn*, 65 and 161–162; and Virilio, *Administration of Fear*, 43.

Chapter 6

Ethics for Modern Thought, Action, and Making

Ellul reframes a traditional Reformed theology of justification in his dialectical writings on technology to expose the self-justificatory momentum of *technological* discourse and to create time and space for an immediate hearing of God's justifying Word in the modern world. Similar theological impulses lead Virilio to emphasize senses and mediation in order to show people what they would rather not sense in technology. Virilio's aim is to create time and space for Jesus Christ to mediate peoples' senses of God, other people, and creation. This chapter draws out the most helpful, dynamic, and enduring constructive insights in the writings of Ellul and Virilio for thought (*epistēmē*), action (*poiēsis*), and making (*tékhnē*) in the modern world, all of which flow from their theologies of justification and mediation. It also considers the individual and collective value of both authors' work for today and move beyond the ethics examined and derived from both authors' writings.

EXISTENTIALLY NAKED PRAYER

For Ellul, confession of nakedness before God as a consequence of justification received from Jesus Christ is the first ethical action (*poiēsis*) to be taken on a daily basis. Choosing the enveloping power of technologies over the clothing graciously given by God must be repeatedly renounced. The work of renouncing power to receive God's gifts every day implicitly and explicitly prepares one to resist all destructive power in the varied forms it exists in the modern world.

One of the most striking details in *Théologie et Technique* is Ellul's apparently passing reference to polymath scientist, anthropologist, and philosopher

André Leroi-Gourhan during a discussion of the postlapsarian states of Adam, Eve, and their sons Cain and Abel. Ellul writes:

> The [fallen] world has become hostile through the powers of aggression that it harbours against man—the ferocious animal, the snake first. Man is now obliged to defend against what attacks him, and so other means are necessary here—weapons, for example (for, why should we reserve Technique for tools? The weapon is the sign of Technique at least as original!) but more so all Technique. A good specialist [on] these questions could write that Technique was a "protective envelope that man built around himself!" (Leroi-Gourhan), and it is true that Technique is at least as much a set of means of protection as [a] means of domination. But, in one case as in the other, what characterises this instrument is its effectiveness.[1]

Ellul's reference to Leroi-Gourhan recalls the history of *technique* at the beginning of *Technological Society*:

> Magic clearly displays the characteristics of primitive technique, as Leroi-Gourhan indicates when he says that technique is a cloak for man, a kind of cosmic vestment. In his conflict with matter, in his struggle to survive, man interposes an intermediary agency between himself and his environment, and this agency has a twofold function. It is a means of protection and defense: alone man is too weak to defend himself. It is also a means of assimilation: through technique, man is able to utilise to his profit powers that are alien or hostile. He is able to manipulate his surroundings so that they are no longer merely his surroundings but become a factor of equilibrium and of profit to him. Thus, as a result of technique, man transforms his adversaries into allies.[2]

The text to which Ellul refers in both quotes is Leroi-Gourhan's *Evolution et techniques*, *Milieu et techniques*.[3] He does not provide exact references but the sections "Le Vêtement [Clothing]" and "L'Habitat [Housing]" in *Milieu et techniques* are the likely source texts for his comments on clothing and *technique*.[4] In the latter section, Leroi-Gourhan writes "the dwelling completely covers the political, linguistic, and clothing area. For clothing, we have already seen that the substitution of one type for another is imposed by conquest: the effective presence of the conqueror is necessary so that fashion turns to forms absolutely different and sometimes less convenient. It is the same, with less speed, for the house."[5] Putting this in theological terms: after the Fall, God conquers the problem of postlapsarian nakedness by killing an animal—a sacrifice—which makes a kind of clothing for Adam and Eve much more durable and protective than their miserable fig leaves.[6] Resonances with *Meaning of the City* and *Technological Society* are strong here (see chapter 4), but in terms of Ellul's ethics, the essential details to draw out in this section are the historical, sociological, and theological connections Ellul makes between postlapsarian *technique*, nakedness, clothing, protection, and conquest. Essentially, Ellul reads Adam and Eve's postlapsarian clothing allegorically, as a theological representation of the material means people use to hide themselves from God. As these materials means include technologies, Ellul's allegorical reading of postlapsarian nakedness and clothing frames his ethics of technology.

The corresponding passages in *Technological Society* and *Théologie et Technique*, with twenty years between them, represent the stability of Ellul's reading of the Fallen human as naked because he or she no longer has a "cloak," "cosmic vestment," or "protective envelope" for protection. Such a person is not only physically naked, however. They also are naked spiritually. In the fullest sense, they are naked existentially.[7] As a consequence, postlapsarian nakedness causes a person to make and to use technologies as he or she works to survive in a threatening environment, to produce clothing, shelter, protection, and food. Of course, this is necessary after the Fall. But it is not a sign of freedom. Rather, it is a "sign of our fallen nature."[8] Genesis 3:17–19 is clear on this truth. To cover up one's postlapsarian nakedness with works and means, and to call that covering free and freedom-proffering, is to justify works and means, which would not have been needed if the Fall had never happened. It also is to expose oneself to the seductive powers of The Spirit of Power and The Spirit of Lies. Whom God's Son sets free is free indeed—not the one freed by postlapsarian works and means. John 8:31–34 is clear on that truth.

The imagery of postlapsarian nakedness and clothing connects to what Paul implores in Galatians 3:26–27: "Now that faith has come, we are no longer subject to a disciplinarian, for in Christ Jesus you are all children of God through faith. As many of you as were baptized into Christ have clothed yourselves with Christ." The theological and ethical point is not that wearing or making material clothing, and what that wearing and making represents in respect to technologies, is categorically sinful. Rather, after the Fall, the wearing and making of material clothing is a necessity, not to be justified as freedom. In contrast, being baptized into and being clothed with Christ is the true mark of a person's freedom. As Paul says in Romans 13:14, "Put on the Lord Jesus Christ, and make no provision for the flesh, to gratify its desires."[9] Putting on, being clothed by, Jesus Christ is the first, and most essential, clothing. All other forms of clothing follow, or do not follow, as we live in faith, knowing, and believing that we are not subject to the disciplinary forces of fallen forms of clothing, but are clothed as children of God through faith.

Returning to Ellul's reading of the Fall in *Humiliation of the Word*, postlapsarian nudity's presence in his theology and ethics stands out.[10] Ellul writes, "[Adam's and Eve's] eyes were opened with respect to what? Their nudity. That is the great first discovery. And that is all. All their new sight allows them to discover and learn is this nudity. And this discovery leads them to make clothing for themselves."[11] Scripturally speaking, "Nudity is essentially the sign of fragility and weakness," "Clothing is the sign of and search for protection," and, "[The Fall] amounts to the discovery of the terrible reality that humanity without God is nothing. Without him we are without protection, without strength, and we are fragile as a breath."[12] So, "We can certainly say that [the Fall] will be the point of departure for 'science and technique,' since these are essentially means for escaping from this weakness and acquiring protection and power," but "there is nothing grandiose and sublime in this."[13] Either we confess our nakedness and fragility to God or we fashion our own clothing as Adam and Eve did; accuse God and the other person as Adam and Eve did; name and dominate as Adam did; kill, flee, and build as Cain did; justify good and evil in separation from God as Adam, Eve, and Cain did. There is nothing grandiose, sublime, or justified in these acts, but there is a gracious God who responds to a person's contrite confession of nakedness, fragility, fallenness, and need. The spiritual clothing we wear orients us as we undertake to clothe ourselves materially. When Jesus teaches us how to pray, his prayer, and our prayer, opens with: "Father. . ." (Matthew 6:9–15 and Luke 11:1–4). The material needs follow. As such, all necessary ethics of technology must grow from the freedom we receive from God:

> The temptation is to concentrate all man's life upon, and dedicate it to, technological and economic development. The temptation is to say, not that man does not live by bread alone, but that he lives primarily or essentially by bread. That the baker should bake bread and that Jesus should eat it is fair enough. But one is not to try to bring the omnipotence of God into the circuit. It is lawful for man to work and to produce. But he must remember that this is a secondary and relative matter.[14]

The primary matter, from which all other matters flow, is putting on Christ and being fed from the mouth of God. But, one might ask: What does it mean, practically speaking, to choose God's clothing over the coverings of technology? Can I put on Jesus Christ and use technologies that make life efficient? How do I resist the discourse, which justifies technology? I take the first and third questions to be the most serious for theology and ethics. Questions like the second one tend to shift toward situational ethics regarding what disciples of Jesus Christ should and should not do, or in this case, should or should not use. For this reason, privileging questions one and three in what follows implicitly and explicitly answers all three questions because if a house is divided against itself, it cannot stand; if we try to serve two masters, we will eventually choose one master over the other. The house must be unified by God, and our master must be God, if we are to stand a chance at getting a true ethic of technology.

For Ellul, the way to plausibly answer these questions starts with the act (*poiēsis*) of prayer; a constant confession of our nakedness and fragility before God; a frequent admission of our need for the clothing of grace and protection that only God gives; a decision renewed each day to resist, consciously and concretely, the protective envelopes of a world filled with countless technologies offering clothing, protection, works, justification, power, and conquest. In the same way, God graciously clothed Adam and Eve after the Fall, even though they fashioned their own clothing out of fig leaves and branches after discovering they were naked before God and each other, God offers clothing to the people who are naked and fashioning their own fig tree clothing. Just as God offered forgiveness and protection to Cain, God offers the same forgiveness and protection to the people who hide in cities and name and justify their works as good in separation from God.[15] Or, to put it succinctly, becoming conscious of our postlapsarian nakedness means we cannot be anything but people who constantly pray existentially naked prayers before God. As Ellul often says, theology and ethics are about being rather than doing. In this case, it is about being people who pray raw, desperate, prayers because they have been baptized into and clothed by Jesus Christ.

In *Prayer and Modern Man*, Ellul captures the practical challenges of prayer in a world that envelops us with the power and parody of technologies. He writes, "We are concerned [in this book with] the actual situation of contemporary man, in our technical, technicalized society, with the situation of the person who does not want to pray because nothing invites him to do so, for whom everything is a deterrent from it."[16] Here, "I shall not speak about the fact that man does not want to, or cannot, pray because he is a sinner. That has always been true. That is nothing new. But I ask myself, how does the situation look today?"[17] He answers the question from various perspectives throughout the book, but its conclusion, "Prayer as Combat," articulates one of the most compelling answers for Ellul's ethics. He says, "All [radicalism] of behavior, of style of life and of action, can only have the prior rupture of prayer as its source. Precisely because our society is given over entirely to action, the person who retires to his [or her] room to pray is the true radical. Everything will flow from that."[18] In relation to thought, action, and making, and how each of us are or are not determined by technologies, this forces an either/or choice; a theological, ethical, and practical choice between either the efficient action of technology or the efficacious action of Jesus Christ.

On the one hand, *technique*, *technology*, and *propaganda* seem to guarantee efficient action and each may or may not be able to deliver on the guarantee depending on the time and place. But what are the short- and long-term consequences of choosing this way of efficient action? On the other hand, God's action, God's action for, with, and in us always is efficacious, but in ways that often resist our often limited, and frankly fallen, understanding of what constitutes *efficiency* and *efficacy*.[19] One of the many gracious, gratuitous, gifts that come with being clothed by Jesus Christ is we are efficaciously freed from the determinations and dangers of the kind of efficiency Ellul describes in his writings on technologies. It is in prayer, existentially naked prayer, that we are clothed and efficaciously freed by Jesus Christ to be who we are called to be and do what we are called to do today. For Ellul, this is an essential place to start when answering the questions asked earlier. As he says, "The person who retires to his [or her] room to pray is the true radical. Everything will flow from that."[20] True, this may seem to be anything but radical. But, Ellul's perspective differs. He believes the radical fight of the Christian takes place in the prayers that wage war on the principalities and powers of our age.

In *Judgment of Jonah*, one of Ellul's first texts, he describes the action of prayer taken by Jonah, as he cries to God from the dark and acidic abyss of a sea creature's stomach. What Ellul writes captures the form of an existentially naked prayer:

> The cry which God hears comes from the depths of the abyss, from sickness and suffering, from the heart which is humbled, bruised, and despairing. This is the cry which produces conversion because things cannot stay as they are, and conversion is a change of route for man. The moment a man decides to change his style of life in this way, the moment he remembers God again, his way which was plunging more and more deeply into the dark is suddenly directed to the light in a dizzy reascent. The truth is that God responds, not to our better feelings, but to the desperate cry of a man who has no other help but God.[21]

But, Ellul also warns,

> When man has somewhere to turn he does not pray to God and God does not come to him. As long as man can invent hopes and methods, he naturally suffers from the pretension that he can solve his own problem. He invents technical instruments, the state, society, money, and science. He invents idols, magic, philosophy, spiritualism, and all these things give him hope in himself that he can direct his own life and control his destiny. They all cause him to turn his back on God. As long as there is a glimmer of confidence in these means man prefers to stake his life on them rather than handing it over to God.[22]

The overflow of works and means in the modern world—including technologies—always tempts us to choose other actions, other clothing, other shelter, other tools, other protection, other weapons, other provision, other conquests, other powers, and other justifications; other than a contrite cry to the one God who pulls a person out of the pitch dark abyss of fallenness and nakedness and flings a person into the brilliant light of grace and protection. "Prayer only has meaning when it is an offering of the man who stands naked before God with nothing in his hands" Ellul writes in *Ethics of Freedom*.[23] Or, *Hope in Time of Abandonment*, "Hope is power and action only in the presence of naked reality."[24] The quality, strength, and vitality of an ethic of technologies depends upon a willingness or lack thereof to confess to God with empty hands on a daily basis that *we too* are naked, that *we too* choose to clothe ourselves with the protection, conquest, power, and works of technologies—and we justify it!—rather than asking, in naked hope and empty-handed faith, for the gracious protection, provision, and justification of God. As Ellul confesses, so ought we to confess: "The man of our times does not know how to pray; but much more than that, he has neither the desire nor the need to do so. He does not find the deep source of prayer within himself. I am acquainted with this man. I know him well. It is myself."[25] Pouring out from these contrite confessions in prayer all other actions will follow, because it is a confessional act of prayer "addressed to the One who issued the call in the first place."[26] As this section has examined the form of existentially naked prayer, the next section will examine its content, and it will continue to answer the questions asked earlier.

A SERIES OF QUESTIONS

It could seem daily confession of nakedness before God is initiated by a person's thoughts and actions. But Ellul believes God the Father already has spoken and is speaking through God's Son and God's Spirit. Three scriptural questions spatially and temporally bring this speech into the present for Ellul, and they communicate, theologically and ethically speaking, the only true source of justification for a person's thought (*epistēmē*), action (*poiēsis*), and making (*tékhnē*) in the world.

From his early to late writings Ellul upends common conceptions of scripture, arguing that while people tend to bring questions to scripture, scripture brings revelatory questions to people, making the person responsible to and for what they read and hear.[27] As he writes in *Living Faith*, "In all [this] babbling of ours we miss the core of biblical revelation when *it* speaks to our faith and asks questions of *us*."[28] What God reveals of Godself in scripture, Ellul argues, "Asks us a question—a series of questions—that is; it makes people responsible (obliged to respond) and throws them back upon their freedom."[29] Although a personal crisis in the 1970s caused Ellul to write "God is silent" in

Hope in Time of Abandonment, there are important and preferable trajectories in Ellul's writings. His descriptions of the fundamental questions asked of people by God offer vital, Christologically anchored, questions, which can harbor theology and ethics even in difficult, painful, and seemingly hopeless situations, precisely the kind Ellul underwent in the 1970s.[30] What follows, then, is a constructive reading of those strands of Ellul's work that aim to offer a word of hope to a person, any person, who is thrown into a moment when they seem to have lost it.

There are many places in Ellul's writing where the fundamental questions of God's self-revelation in scripture appear, but *Ethics of Freedom* and *Living Faith* provide the most extensive commentary on these themes. In fact, in *Living Faith*, Ellul finds "questions [which] run all the way through the Bible, with three high points."[31] He writes:

> The first question asked of us is a double one: "Adam, where are you?" and "Cain, what have you done to your brother?" In other words, Adam, the perfect communion between us has been broken; where have you gone astray? What have you become? Or, as they say nowadays, where are you coming from? And you, Cain, like Adam you were a guardian, a protector; what have you done to the one closest to you?
>
> The second question, at the heart of our discussion, is the one Jesus poses to the disciples: "And you, who do you say that I am?" He doesn't proclaim himself, he makes no declaration; instead it's up to us to discern this. God in Jesus limits himself to asking the question—the real question, the only important one: whether we are alone on the earth; not whether "there exists a God hovering in the clouds," but where are we met "where we're at" by God himself.
>
> And the third, in the last days, after the resurrection of Jesus Christ: "Whom do you seek?" and then, "If it is my will, what is that to you?" These are questions about the reality of our search—whom *are* we seeking, after all? What are we searching for? Where is it all going, this human history where we cannot recognize the figure of the Crucified-but-Risen One? And at the same time we have a decision to make: Jesus says "If I will . . . this or that, what does it matter to you?" What really *is* important to us, without our pretentious little claim to knowledge, science, judgment.
>
> All through scripture these three questions serve as a framework for the questions that God asks us and that faith is called to answer.[32]

God the Father has acted and has spoken; is acting and is speaking; will act and will speak in and through His Son, Jesus Christ, the Crucified-but-Risen One. The importance of the scriptures is that, as we read and as we hear *what* is revealed in and by the scriptures, we encounter, again and again, *who* is revealed in and by the scriptures. The three questions asked earlier in these scriptures come to us who have put on Christ, as we encounter Christ, who animates our lives, in and outside the church, as each question bursts into the present, sparks responsive action (*poiēsis*), and inspires confessions of nakedness and need for a God who reveals Godself in Jesus Christ the Crucified-but-Risen One. Ellul writes, "Prayer, in the degree [to] which it is the decision we have described, addressed to the One who issued the call in the first place, the Wholly Other and at the same time the Transcendent, is in reality the exact counterpoint of the rigorous mechanism of the technological society."[33] If we do not have the content for the form of existentially naked prayer, scripture gives us the content, and the questions revealed in scripture, which Ellul highlights and comments on earlier are a theologically, ethically, and practically critical place to start. Furthermore, when we have no words of our own; when we feel like groaning in frustration or despair; and, when it seems as if we cannot read nor hear the truth of the scriptures today, the Holy Spirit intercedes on our behalf, giving us the true words and groans, illuminating the scriptures, so that we can read and hear them in a dazzling light today.[34] Viewing prayer from this perspective, Ellul presses us to conclude that, when we pray, we ought to speak less of our words, and hear more of God's words spoken to us and for us, working in us and for us.

The spatiotemporal push of the content of these questions demands emphasis because each question reveals previous acts of God to us which act upon us in the here and now. *Where are you? What have you done? You, who do you say that I am? Whom do you seek? If it is my will, what is that to you?* All of us are asked these questions by God in the here and now, and how we answer these questions in the here and now, will establish our justification, in the here and now. But, and this is Ellul's warning, it is difficult, very difficult indeed, to hear clearly and to respond concretely to these questions in the here and now if we are enveloped by powerful technologies and seduced by The Spirit of Power and The Spirit of Lies. "Our first task is to listen and not to act" Ellul writes in *Meaning of the City*.[35] The first act of prayer always is preceded by God's action as one listens for and to God—which is an act!—and from this listening for and to God, a here and now response to God's questions explodes into our lives. This will obliterate self-justification, and spark actions; grateful, obedient, responsible, gratuitous, free, sundry, spontaneous, actions that glorify and bear witness to a God who comes to us in and through God's Word, Jesus Christ, the Crucified-but- Risen One. If any one of us lacks practical wisdom for how this is done today, one only needs to ask God for the wisdom, today. "It is precisely the wise man who knows that he is not wise, and that unless God gives him the knowledge of the good in his revelation, he knows nothing, and is living in folly."[36] Any ethic, both in relation to and beyond technologies, must be fed by these questions, and this wisdom, graciously and freely held out to us by God, to us and for us.

As the previous section focused on being clothed by Jesus Christ, this section focuses on being fed by Jesus Christ. Ellul writes in *Hope in Time of Abandonment*, "The hope of Jesus Christ is never a dash of pepper or a spoonful of mustard."[37] The hope of Jesus Christ is "bread and wine, the essential and basic food itself, without which there is only the delirium of knowledge and an illusion of action."[38] As Jesus replied to The Spirit of Power and The Spirit of Lies within the wilderness, so ought Christians and church reply.[39] Even if technologies were able to turn stones into bread, disciples and the church live by the bread and the wine given by God.[40] Hearing and responding to these questions today, asking for God's wisdom today, are the first acts, which clothe and feed the disciple; make ready the disciple and the church for all thought (*epistēmē*), action (*poiēsis*), and making (*tékhnē*) in the modern world.

A MATTER OF PROSAIC DECISIONS

Ellul's sociology, theology, and ethics tend to focus on the individual, but Ellul also wrote about the sociological phenomenon of the church and how contingent ethics can be formed in churches in ways, which respond responsibly to God's Word while resisting rigid legalism as well as flaky liberalism. For Ellul, the Holy Spirit's presence and action in the church makes ethics a matter of prosaic, but dynamic, decisions for the believer and for the church, but only occurs when believers are listening to and speaking with God and each other.

In *Theological Foundation of Law*, Ellul describes what he believes the church ought to be and do in the world.[41] First, the church is founded "by the word," "for the sake of the word," and "must be granted the fundamental freedom to proclaim the word of God. God sends the church into the world mainly for this purpose."[42] Second, "To proclaim the gospel is not only to proclaim the good news of the forgiveness of sin. It is also to proclaim the concrete consequences of this good news. It is to announce that Jesus Christ is Lord of creation, with all the implications this entails."[43] Third, the church must not separate the "preaching of the cross" from "the prophecy of the parousia," thereby assuming the role of a "watchman" for "the earthly city."[44] Concretely speaking, the church "[suffers] the same afflictions which Christ suffered for mankind" with compassion and service for others.[45] Fourth, and finally, "Only when Christians fully understand their role and their responsibility in society and in matters of law, and are keenly aware of the power at work in them through the Holy Spirit, will the church be in a position to speak and to act as it is required to do."[46] For Ellul, the Holy Spirit's presence and action in believers' lives makes possible the church's hearing and proclamation of God's Word, concretized in the acts of preaching, prophecy, teaching, and compassionate service. Could the same be said for the formation of ethics in the church? Not explicitly in *Theological Foundation of Law*, but explicitly in *To Will and to Do*.[47]

Showing an affinity with Luther in *To Will and to Do*, Ellul argues that, even in the best of circumstances, a person is simultaneously justified and sinful between the Fall and the Eschaton. As he puts it, even the person who is redeemed by Christ is, "*Simul peccator et iustus*. [So,] the Christian has not become better able to do the good."[48] Being at once justified and sinner liberates rather than oppresses, however, because it launches the justified sinner toward "the necessity for God's intervening to guide our lives [putting] an end to our pretending to erect a Christian morality," revealing to us that "we can bring to light the commandment of God as much as we please, [but] if the Holy Spirit is not acting within us our knowledge is in vain. We shall never be able to carry it out."[49] Then, "[The] intervention of the Holy Spirit alone renders us apt for hearing and obeying the commandment of God. [The Holy Spirit] alone makes us hear the commandment as really addressed to us personally, and in so doing [gives] us the power, the capacity, for carrying it out."[50] Consequences for ethics follow: "The innovation of the faith is a ceaseless invention of new ethical forms," a "response given by man to the confrontation between the Holy Spirit's requirements and the concrete situation in which he finds himself," making ethics "a matter of prosaic decisions, of microscopic adjustments, which cannot (because of this characteristic) be the object of any blueprint, of any prior decision. Otherwise we would lapse into casuistry."[51] Believers are drawn together in the church by the Holy Spirit who intervenes to enable believers to hear God's Word and to make it possible for believers to actualize their responses in word and action. Prosaic ethics are the surplus of believers' actualized responses to the intervention of the Holy Spirit and the command of God's Word both in and outside the church.

It may seem that making ethics a matter of prosaic decisions risks churches and disciples sliding into disorganization, inconsistency, and vice, but Ellul is convinced that God's consistency makes up for creaturely inconsistency. He writes, "God is the same, yesterday, today, and forever" and "[God] at no time deceives us about himself. That which he revealed yesterday is always completely valid. Therefore we can count on a steadfastness, on a continuity on the part of God."[52] So, "It is legitimate, even ordained, that we should lay hold of [past] revelation and seek in it an ethical instruction for today. And for that reason we should be well assured that there is no conflict between the objective revelation *hic et nunc*, between scripture and the Holy Spirit, between the permanent will of God and his will *hic et nunc* for each [person]."[53] Believers and the churches in which believers gather are bound to the history of all believers and all churches, and as such, they are responsible to and for God's revelation in the past just as they are responsible to and for God's revelation in the present. Consequently, a believer or a church cannot, at least in good faith, justify disorganization, inconsistency, and vice with an appeal to God's consistent and steadfast revelation in the past or the present. Prosaic ethics, on the other hand, constitutes a church's flexible, Holy Spirit animated, responsibility to and for God's consistent and steadfast action in the stream of history. On this point, Ellul writes:

> Ethics is essentially fluid at that juncture between the word of the Lord and the stream of circumstances, of social, political, and economic relations; once it is recognized that ethics can only be valid for the perhaps brief period during which things are stabilized, and that it should be modified in accordance with events, then we perceive the possibility—not the necessity nor the constant renewal, but only the possibility—of a coming together of the *hic et nunc* of the word of God and the *hic et nunc* of a valid and true ethics for man. For the current application of the word of God, which is too often interpreted as being only for the soul of the believer who receives that particular revelation, is valid as well for the church, and for that reason it contains an element of duration, a registration in the course of history, and is not merely a flash of lightning in the heart's dark night.[54]

The meeting of God's Word with believers and churches in the past and the present always resonates in the course of history. This is what happens when believers gather to listen to and to speak with God and with each other, articulating such listening and speaking in prayer, preaching, prophesying, singing, and teaching—in all ecclesial listening and speaking!—animated by the Holy Spirit's presence and action. If such ecclesial listening and speaking does not take place in the church, then churches become filled with noise, clanging cymbals, and in the modern world, *technique*, *technology*, and *propaganda*. From there, ethics becomes a flash of light in the dark, and slides into disorganization, inconsistency, and vice occur.

In *Ethics of Freedom*, Ellul brings all of this to bear on the leaders and the laity of the church. There, he observes as a consequence of "the more or less complete break between work and vocation under the pressure of capitalism, mechanization, and technologization," there is "a crisis of vocation among Christians."[55] He describes the crisis as follows:

> Vocation is contested by Christians because those who bear responsibility in the church, e.g., pastors, feel that they are undervalued in a technological world in virtue of the fact that they themselves are not specialists and especially technicians. Obeying a vocation and therefore preaching, directing a congregation,

> and engaging in the [cure/counseling] of souls can seem to lack seriousness in a world of engineers and producers. The result is that these shamefaced pastors want to become technicians themselves. They take up psychoanalysis, group dynamics, social psychology, and communications. It is as psychoanalysts and so forth that they want to act in the Christian community and no longer because they are called by God to his service. [But] it is an illusion to think that technology may be appended to a vocation or will serve it. In fact it becomes a substitute for it.[56]

There *is* extraordinary pressure for pastors to be specialists and technicians, engineers and producers, psychologists, and statisticians. Such pressure burdens and bends if it does not break the vocation of the pastor. For the laity to pressure, a pastor to be anything but a pastor is to obstruct the "coming together of the *hic et nunc* of the word of God and the *hic et nunc* of a valid and true ethics for [people]," which the pastor is called to cultivate in the church.[57] For a pastor to want to be a specialist and technician, engineer and producer, psychologist and statistician, rather than to want to be a pastor, is an equal, if not greater, obstruction to the church's listening and response to God's Word in the present. The same must be said for all people leading and serving in the church, including, but not exclusive to, pastors, apostles, prophets, evangelists, and teachers, as they equip the church's members for ministry and tend to its health.[58] Note especially the auditory connotation of *vocation* here. Hearing and responding to God's vocation, the vocalizing and hearing of God's Word in the presence and action of the Holy Spirit, can be blocked by the discourse of *technology* and the manipulations of *propaganda*; can be hindered by congregations who demand technicians and engineers; and can be occluded by pastors and leaders who have become technicians and engineers. As for the laity, Ellul writes:

> The church, too, exists at the organizational and social level. It is not just a gathering of believers around the Lord's table. If it is in symbiosis with the world, this is not because it merges into it. It is because the laymen, living in the world and experiencing within themselves the tension of twofold allegiance, come out of the world into the church and then go back to the world. To achieve merger the church does not have to wither away. It has to exist in every possible visible dimension in order to make sure that it embraces every possible layman and is not just a bloodless fellowship. This is an indispensable condition for the layman if he is to carry the freedom which is given to him and inscribe the incarnate presence of the Lord in decisions which do not conform to the schemata of society. It is on the basis of a church which is a strong body and community that this is possible for the layman.[59]

The laity is thrown into a threatening world thick with *technique*, *technology*, and *propaganda*, so the laity must go into the world prepared to resist it with support from the church and must return to the church anticipating a gathering of believers free from the machinations and power of *technique*, *technology*, and *propaganda*. Practically speaking, the laity encounters challenges in the world, which requires pastors, leaders, and laities to *listen* to and *speak* with each other and to not *manipulate* and *pressure* each other as they work together to collectively hear and to respond to God's gracious and liberating Word in, to, and for the church, which is drawn together and animated by the Holy Spirit. In sum, the more mediations of technologies are brought into the institutional church, the less likely institutional churches will be able to hear and to respond immediately to God and neighbor: "Technology has become a mediation in all actions and for all purposes," and "the regularity of effects, the exclusiveness of means, and generalised mediation leave no place at all for the concept of vocation."[60] Ellul is suspicious of the mediations of technologies because they can silence the auditory essence of vocation heard and responded to by believers in and outside the church.

Essentially, all of the aforementioned contexts amount to attentive-to-God-and-neighbor decisions, prosaic decisions, made by all the people who constitute the institution of the church, pastors, leaders, and laity; communal, prosaic decisions, for freedom from rather than capitulation to the determinations of any this-worldly institution. Ellul grants that freedom is expressed in and through institutions, but he also contends that churches must "fight the institution, not in order to destroy it, but in order to secure some place for freedom and to achieve the salutary effect of inserting tension into the institution itself."[61] Specifically, "The attack should be directed against two specific aspects of the institution, namely, abstraction and bureaucratization" because "the institution is always a power of abstraction. It replaces regard for the objects with regard for the rules which are imposed on it."[62] Resisting the tendency toward the manipulative, hierarchical, abstract, and bureaucratic power of social institutions begins with an insertion of freedom and tension into, an attack on abstraction and bureaucratization in, the institution of the church. It begins with and flows from the collaborated effort of leaders and laity who join together to see a "coming together of the *hic et nunc* of the word of God and the *hic et nunc* of a valid and true ethics for [people]."[63] It begins with a refusal to use *technique*, *technology*, and *propaganda* to manipulate or pressure anyone in or outside the church. Finally, it begins with a resistance to the principalities and powers, including The Spirit of Power and The Spirit of Lies, at work in the hostile world today. For, "Wherever we find the worship of power—and this is a common phenomenon today—the church should be ready to attack."[64] God's "[Revelation] calls power into question rather than setting it on a throne."[65] The more churches, actual institutional churches today, embrace *technique*, *technology*, and *propaganda*, the less likely they will be able to hear and to respond to their vocations given from God, the less likely they will be able to form prosaic ethics, which foster freedom, and the more likely they will be to set power on a throne.

The recently discovered and published second volume of *To Will and to Do* summarizes succinctly how believers and churches can think (*epistēmē*), act (*poiēsis*), and make (*tékhnē*) within the pulls of sociological necessity and the pushes of revelation as they have been described thus far. Here, Ellul writes of the formation of ethics as they are lived prosaically within the church:

> Ethics has a double character. On the one hand, like any human group, the Church will provoke, produce, "secrete" a morality: it cannot escape this sociological necessity. But as people of God, and before the Revelation contained in Scripture and under the inspiration of the Holy Spirit, [the church] is also called to formulate an ethic. The duality reproduces exactly the duality for the Christian: on the one hand as a sinful man . . . he decides the good and evil—but on the other hand, as having received the Revelation, and animated by the Holy Spirit, he knows another Good and is called to live according to a new ethic. Only, he is never completely stripped of the old man, and the ethics he can formulate will always be a reminder of his position as a natural man. Thus, this communal ethic is necessarily understood in this tension between the sociological character of the Church and the inspiration of the Holy Spirit. [The tension] includes both. And it cannot be said that this is bad—otherwise, the Ethics formulated would be unlivable and [outside] of time: it would no longer be an ethics.[66]

For prosaic ethics to be formulated in the lives of believers and churches, the people who have the postlapsarian ability to decide good and evil, but have come to know another good through God's self-revelation, must gather in the church to form prosaic ethics with other people who also have the postlapsarian ability to decide good and evil, but have come to know another good through God's self-revelation. *Technique*, *technology*, and *propaganda*, institutional abstraction and bureaucratization, press in on pastors and leaders to become technicians and engineers, blocking and breaking their vocations. Desires for social power and spiritual power only obstruct rather than cultivate the freedom, growth, worship, and witness that should characterize such free and gracious gatherings.

EITHER SINCERITY OR INSINCERITY

Ellul avoids making decisive claims about universal salvation in his writing from the 1940s to the 1970s. The 1970s to the 1990s see the inconclusive nature of his statements on this topic harden. This hardening alters his account of which postlapsarian parts of history and works will be recapitulated at the Eschaton. The implications of this doctrinal shift for Ellul's ethics are examined in this section and a constructive response is offered here.

In *Meaning of the City*, written in the 1940s but not published until 1970, Ellul leaves the question of universal salvation and recapitulation—and thus the dialectic of history—open. He writes, "It is in Jesus Christ that God adopts man's work. For Ephesians 1:10, translated literally, tells us that Jesus is the great recapitulator."[67] This "includes things invented by man, what [man] laboriously put together piece by piece learning from experience and failure. Both his technical failures and the marvels of his cleverness. All this is 'recapitulated' in Christ, summed up in him, taken over by him."[68] But, Ellul clarifies, "God's way of judging is not ours. What will he preserve? We have no way of telling."[69] Five years later, Ellul's commentary on the Book of Revelation *Apocalypse* is published, and he writes, "We have this remarkable double affirmation: on the one hand, the celestial Jerusalem is a creation of the absolutely new [creative act] of God, without common measure with anything else; but on the other hand, it is the perfect synthesis, made by God, of all history, of the total life of humanity, of all man's works, all his creations, all his ventures: it is the recapitulation."[70] The difference between these two passages, and the changes in Ellul's theology they represent, produces an either/or decision for how Ellul is read before and after the 1970s. Either the shift in Ellul's theology actually occurred and he wrote sincerely about universal salvation and recapitulation after the 1970s, or after the 1970s, Ellul was emulating Kierkegaard and writing insincerely about universal salvation and recapitulation to provoke a *yes! no!* or *could it be possible?* reaction. Either way, reactions are provoked.

Just as the previous section opted for the more constructive trajectories of Ellul's writings, this section responds to this either/or question with the aspects of Ellul's work that leave his questions about universal salvation and recapitulation open. Specifically, in response to the question "What will [God] preserve?" this section and book answers with Ellul's early writing that "we have no way of telling."[71] But it also answers with the Ellul's later writing, "Not only is the whole meaning of history rooted in the paschal Mystery, but the paschal Mystery is that of liberation."[72] Contrary to Ellul's impulse to penetrate the paschal mystery of liberation in his later writings, I suggest that the facets of his work, which emphasize the mystery of salvation and recapitulation—essentially, our inability to judge who and what will be redeemed at the Eschaton—stand as the more constructive theological and ethical impulses. This is because the latter emphasizes, theologically and ethically, the truth that all salvation and all good works are gifts from God. Accordingly, as Ellul writes, "God takes the initiative. That is to say, he begins by giving, and the rest follows."[73] Furthermore, in *Hope in Time of Abandonment*, an important text that marks the tectonic shift in Ellul's theology, he writes earnestly and with an apocalyptic hope:

> If we have the idea that the world develops through the wonderful works of man, that it goes from progress to progress in that way toward the Kingdom of God (and, at best, only by political and social revolutions!), that there is an unbroken continuity between this world and the Kingdom, that the way is prepared for the latter through political, technological, and scientific action, then one is dealing with a motionless object-God, a God who is no longer a stopgap, but a porcelain vase set in a corner: "Wait for us. We'll take care of it." The Apocalypse is tied to the thought of a God who intervenes in history, who makes his own decisions and acts as sovereign, creating the world he wants through his almighty Word, whose fiery approach melts mountains and causes man and his works to collapse. It is to take the living God seriously.[74]

The apocalyptic intervention of God in history collapses the works of humanity—including technology—and creates the world *God wants*. For that reason, I propose to read Ellul's ethics for thought (*epistēmē*), action (*poiēsis*), and making (*tékhnē*) in ways that are theologically anchored in the sovereign God who makes God's own decisions, intervening to gift God's people with thoughts (*epistēmē*), actions (*poiēsis*), and works (*tékhnē*), which take the living God seriously. Accordingly, I prefer the strands of Ellul's writings that emphasize the theological and ethical point that, if believers and churches are to have any hope, it will be a hope rooted in a desire for God to intervene apocalyptically and create the world God wants to create; through us partially perhaps, but ultimately, and finally, by God on the last day. But until then, as Ellul so often says, ethical action today ought to rely less on doing *this* or *that*, and more on *being present* in the world with God and with others.[75]

AFTER ICONOCLASM

Ellul's use of a modernized Reformed iconoclasm, desacralization, and demythologization in his writings on technology offers vital resources for theology and ethics today, but they also thin out what could be a thicker account of creation's mediative capacity and how it shapes thought (*epistēmē*), action (*poiēsis*), and making (*tékhnē*). Virilio thickens these thinner aspects of Ellul's ethics of technologies in three related ways.

First, Virilio's frequent engagement with Dietrich Bonhoeffer's critique of immediacy turns readers to a positive account of Jesus Christ's mediation. Second, Virilio uses Bonhoeffer's theology of mediation to develop an ethics of rest and silence within creation, which creates time and space for the light (*phōs*) of Jesus Christ to saturate a person's senses as he or she thinks (*epistēmē*), acts (*poiēsis*), and makes (*tékhnē*) in the world. Third, Virilio's bunker ecclesiology aesthetically concretizes his theology of mediation and his ethics of rest in the form of a church he designed with Claude Parent, modeling a way for believers and churches to transform horrible habitats of war, terror, and death into habitats of peace, rest, and life. This section will focus on the first element of Virilio's work on technology, which reinforces Ellul's.

Virilio's theology of mediation, and how it frames his writings on technologies, emerges when his refrain in interviews that negativity is a positive task is joined to his frequent quoting of Dietrich Bonhoeffer's critiques of immediacy. Virilio has been criticized for being too negative and for not providing constructive solutions to the problems he exposes.[76] He often responds to these criticisms with the paradoxical statement, "Negativity is a positive task," which, as suggested earlier, offers a hermeneutic for the negative imagery in his writing.[77] For example, in *Lost Dimension*, Virilio writes:

> Bonhoeffer once wrote: "In its form, creation faced God in an altogether new manner, and in this face-to-face encounter, creation belonged to [God] completely." Now—with the optoelectronic ubiquity and its incidence over the configuration of territory—we can revise this statement, as regards the teletopological metamorphoses of the city: "In its form-image, creation faces man in an altogether new manner, and, in this interface, creation appears to belong to *him* completely." It must go without saying that this term "completely" is an optical illusion, a simultaneous energy-and film-derived illusion.[78]

There are glimpses of positivity, even positive theology, in the reference to Bonhoeffer's Creation and Fall in this passage. Virilio uses the contrast of positive/negative theological imagery to delineate pre and postlapsarian creation. On the one hand, creation face-to-face with God before the Fall; on the other hand, creation turned away from God with humanity turned inward—an allusion to Bonhoeffer's "God and humankind-imago-dei versus god and humankind-sicut-deus"—producing a postlapsarian enveloping of humanity in the energetic (*kīnēsis*) illusion that creation belongs to humanity rather than God. The technical terms Virilio uses here—"optoelectronic ubiquity" and "tele-topological metamorphoses"—are meant to illustrate how far people have moved away from God and creation and turned inward after the Fall. Such terms help capture how vision enhancing and expanding technologies are oozing over the earth, morphing the appearances of reality, and constructing energetic illusions of control and domination. So it is overly technical terminology used to form negative images of powerful technologies altering how we view ourselves and the world during an overly technical historical moment. But, at the same time, readers who follow Virilio's lead to *Creation and Fall* will discover the following positive theology in Bonhoeffer's writing:

> The divine praise that God prepares out of the rude darkness of the unformed is to be completed through its being given form. The creation still rests entirely in God's hand, in God's power; it has no being of its own. Yet the praise of the Creator is complete only when the creature receives its own being from God and praises God by its own being. In the creation of form the Creator denies the Creator's own self, in that this grants form to what is created and grants to it its own being or existence before the Creator; in that the existence of what is created serves the Creator, however, the Creator chooses to be glorified. Thereby the Creator enormously increases the power of the creation, by giving to creation its own being as that which has form. In this form creation exists over against God in a new way, and in existing over against God it wholly belongs to God.[79]

As suggested in chapter 1, Virilio might have said in interviews, "I have always been utterly unable to write about my faith [because] I do not have the gift for [it]," but that does not mean he did not use theology, nor refer to theologians, in his work.[80] Moreover, what Virilio does have a gift for is forming negative images of positive theology, such as the negative image he forms of Bonhoeffer's positive descriptions of Genesis 1:1–2 in the quote earlier. Accordingly, by encountering Virilio's negative imagery of post-lapsarian creation, coupled with a glimpse of a positive image of prelapsarian creation—"when the creature receives its own being from God and praises God by its own being"—might not readers be turned to the creative God gestured to in this quote? Indeed, with such negative imagery, Virilio pushes readers to make the positive turn—especially when, for example, he refrains "immediacy is a delusion" in reference to Bonhoeffer's *Discipleship*.[81] As Bonhoeffer writes:

> In becoming human, [Jesus] put himself between me and the given circumstances of the world. I cannot go back. He is in the middle. He has deprived those whom he has called of every immediate connection to those given realities. He wants to be the medium; everything should happen only through him. He stands not only between me and God, he also stands between me and the world, between me and other people, and things. He is the mediator, not only between God and human persons, but also between person and person, and between person and reality.[82]

Only then does Bonhoeffer deliver his critique of immediacy:

> So people called by Jesus learn that they had lived an illusion in their relationship to the world. The illusion is immediacy. It has blocked faith and obedience. Now they know that there can be no unmediated relations, even in the most intimate ties of their lives, in their blood ties to father and mother, to children, brothers and sisters, in martial love, in historical responsibilities. Ever since Jesus called, there are no longer natural, historical, or experiential unmediated relationships for his disciples. Christ the mediator stands between son and father, between husband and wife, between individual and nation, whether they can recognize him or not. There is no way from us to others than the path through Christ, his word, and our following him. Immediacy is a delusion.[83]

By frequently referencing Bonhoeffer's negative descriptions of immediacy Virilio is turning readers to Bonhoeffer's positive descriptions of Jesus Christ's mediation. For Virilio, then, just as it is for Bonhoeffer, "[Jesus] is the mediator, not only between God and human persons, but also between person and person, and between person and reality." Consequently, for Virilio, Jesus Christ is the one who ought to mediate a person's sense of speed (*drómos*), movement (*kīnēsis*), and light (*phōs*) in creation (*phusis*), as the person lives in relation to God, other people, and God's creation. For another example of Virilio using negative imagery to turn readers toward positive theology, a shockingly negative image of the Incarnation appears in *City of Panic*, and it makes a similar turn to the turns just highlighted:

> The theology incarnate of Christianity has thus been outstripped by the theology disincarnate of the suicide bombing, in which the crime against humanity presents itself in the eyes of all as the SUMMUM [climax] of the "aesthetics of disappearance" that so profoundly marked the twentieth century, from the invention of the movie camera to Auschwitz and Hiroshima. This time it is the ethics of disappearance that have ravaged the world, a world where speed and morality have never ceased following diametrically opposed curves, from the blasting of the famous "sound barrier" to the acquisition of "escape velocity," that panic-stricken symbol of a world in flight that portends the imminent coming of the catastrophe of CYBERNETIC nihilism, in which the synchronization of individual emotions will finish off the job of the standardization of public opinion, which will then be unanimous.[84]

Apart from forming a visceral Christological indictment of two of some of the most terrible events of the twentieth century, Virilio also forges an important connection between theology, ethics, and aesthetics. The theology incarnate he mentions at the start of the quote confronts the theology disincarnate he describes throughout the quote. A twentieth-century articulation is thus offered of the relationship between the City of God and the city of man. After all, the text is titled *City of Panic*, implying the city of man, in contrast to the City of Peace, the City of God, its positive opposite.[85] In this case, Virilio implores the church to *be* incarnate in relation to the city of man, and never be outstripped by any theology, ethic, or aesthetic that leads to idle watching as concentration camps are filled and atomic bombs are dropped. So, the negative image can be reversed, revealing the positive, constructive, side of Virilio's theology, ethics, and aesthetics: through Jesus Christ's mediation, a theology *incarnate* produces an aesthetic and ethic of *appearance*, which resists tragic aesthetics and ethics of *disincarnation* and *disappearance*.

For a final example, this time Trinitarian, Virilio also forms negative images of the Trinity to affect positive turns in chapter "The Accident of Science" of *Crepuscular Dawn.* Recall here the previous chapter's conclusion that when Virilio says *accident*, *sin* is implied:

> Einstein recognized three bombs: the atomic bomb, the cyber bomb, and the genetic bomb. Alongside Auschwitz and Hiroshima, the atomic bomb sets off the question of the possible end of the human species through the extinction of a way of life, through pollution of the environment, through radiation poisoning. But the information bomb was born at the same time—[the ENIAC computer] and calculators made the atomic bomb possible, and it was the information bomb that allows one to decode the encoding of the human genome map [. . .]. Then they set off the second frenzy, the cyber bomb, the bomb of information [. . .]. The information bomb is in the process of programming the genetic bomb, that is, the modification of the human species, its plurality, its fractalization [. . .]. The three bombs work together in a relationship of all-out war; they reinforce one another. But in the middle of it all, it is the information bomb that is knowledge. It decides.[86]

The negatively rendered perichoretic imagery—"the three bombs work together in a relationship of all-out war; they reinforce one another"—can be reversed for positively rendered perichoretic imagery: Father, Son, and Spirit move together in a relation of peace; they reinforce one another. Likewise, the negatively rendered *logos* imagery—"in the middle of it all, it is the information bomb that is knowledge"—can be reversed for positively rendered Jesus Christ the *Logos* imagery: in the middle is Jesus Christ, the mediator, who is knowledge. Of course, Virilio is not unaware that the first detonation of an atomic bomb was named "The Trinity Test."[87] A photo of the "Trinity Site" monument is featured in *War and Cinema.* Or, in *University of Disaster*, Virilio ask rhetorically, "Following what happened in the Catholic Church in the twentieth century under Karol Wojtyla, its pontiff, shouldn't this retrospective soul searching, which clearly started with Julius Robert Oppenheimer after the success of the 'Trinity Test,' and especially after the bombing of Japan, be revived?"[88] More negatives images used to turn readers toward the positive, thereby providing an illuminating entry into Virilio's theology, ethics, and aesthetics.

It is important to clarify that Virilio's negativity in these passages, as well as others like them, constitutes a form of negative theology (or negative method) rather than a strictly negative theology or style. If it were a full negative theology, he would not offer glimpses of positive theology as he does in the passages examined in this section and throughout his writing. In this sense, embedded within Virilio's form of negative theology, are glimmers of hope, fragments of Messianic time, and ultimately, the positive presence of God's self-revelation in Jesus Christ.[89] Likewise, if it were a negative style, there would be no need for Virilio to reference theologians and theology. Rather, he could just write with negative tone without bringing positive theology into the negativity. So, as Virilio says, "Negativity is a positive task" in his work—especially if, as demonstrated here, the negativity turns readers to the positive presence of Jesus Christ "*the mediator*, not only between God and human persons, but also between person and person, and between person and reality."[90] Conversely, the negative imagery conjured by Virilio renders a brutal representation of what the world and life is like *without* Jesus Christ as mediator, thereby forcing a turn to the negative or the positive. Another way to put this is that, when Virilio writes with such a negative perspective and tone in his work, he is often engaged in a form of the death of God theology, but an assuredly subversive form, in which God has defeated death, and is present with us—whether or not we can perceive it—in the light of Jesus Christ's incarnation, crucifixion, resurrection, ascension, and imminent return.

THE IMPRESSIONIST WAY OF LIFE

Impressionist artists momentarily distanced themselves from modernity's races (*drómos*), movements (*kīnēsis*), and lights (*phōs*) to rest and to be silent in their environment (*phusis*). For Virilio, the impressionist way of life illustrates how a person's senses can be illuminated by Jesus Christ.

Once Virilio's theology of mediation is understood, it becomes easier to explain why fragments and constellations of beautiful, hope-filled imagery appear amid the predominately harrowing and despair-inducing imagery in his writing. For example, in *Negative Horizon*, he alludes to the possibility of God's revelatory truth springing up from the ground in the following way:

> I have always been convinced that everything that will determine the novelty, the originality of tomorrow, is already present in the moment, concealed in the every day vision of each person, which accounts for this constant will to practice

> a perspicacity worthy of my desires for change and renewal. The field of vision always seemed to me comparable to what the ground is for archaeological exploration. To see is to be lying in wait for what must spring up from the ground, nameless; for what presents no interest whatsoever, what is silent will speak, what is closed is going to open, it is always the trivial that is productive, and so this constant interest in the incidental, in the margins of whatever sort, that is, [this constant interest] in the void and the absence.[91]

Could it be that by writing "what is silent will speak, what is closed is going to open" Virilio is responding with a word of hope to Ellul's protest in *Hope in Time of Abandonment* "God is silent" and "we are living through the experience of a closed world"?[92] This is certainly possible, since *Negative Horizon* was published twelve years after *Hope in Time of Abandonment*. However, Virilio makes no reference directly to Ellul in *Negative Horizon*. Either way, Virilio's emphases here form a constructive theological corrective to Ellul's pursuit of, one could even say preoccupation with, an immediate hearing of God's Word in the present. In view of what the previous section and chapters have demonstrated, and what this passage from *Negative Horizon* shows, Virilio's response is that Jesus Christ's mediation causes the silent to speak and the closed to open in creation. In this light, one can find a hope *in* silence and *in* closure, rather than, seek hope *in spite of* silence and *in spite of* closure.

In one of Virilio's earliest essays, he describes an "impressionist way of life," which forms a vision of Jesus Christ's mediation of the believers' senses as they live—as they think (*epistēmē*), act (*poiēsis*), and make (*tékhnē*)—in creation being renewed.[93] For Virilio, "the impressionist way of life" relates to the "unwanted morality, that of a scattered and small in number humanity, restful, as camouflaged in space and time, in a liberty dependent on the unpredictability of its habitat."[94] As referenced in an earlier chapter, the latter passage alludes to Israel, the scattered, small, and seemingly insignificant group of people chosen by God and enabled, through God's self-revelation, to perceive and live within creation in ways that are different from the other nations. Now, reading these passages through a Christological frame, Virilio's "impressionist way of life" assumes its fullest form through Jesus Christ's mediation. In particular, Virilio's blending of Neoplatonism with frequent references to impressionist artists helps to explain what he means by the "impressionist way of life" and how it is Christologically mediated.

For Virilio, the impressionist artist models what it is like to be filled with the light of Jesus Christ, because the impressionist artists let light be *impressed* on them and represented their *impression* of light to the world. In *University of Disaster*, he writes, "Faced with such a profound identity crisis, the subjects of this photosensitive age now want to glitter for all they're worth, to become electro-luminescent. They are no longer content to be both 'impressionists' and profoundly impressed by the shock of the images and the miseries of the times; they want to be luminous, dazzling. Whence the hunger for instant recognition, in real time."[95] Whereas, the impressionist artists were impressed by—they recognized and represented the various appearances of light in their time—many people today want to project their own light, through technologies, in order to be recognized by others. As Virilio suggests, such projections, rather than recognitions and representations of light, are a cause of great shock and misery in the modern world. In *The Great Accelerator*, he makes a similar point, but offers more insight into his understanding of and esteem for the impressionist artists:

> Science, the BIG SCIENCE of physics, is not, in actual fact, at peace with itself. It is at war with time: "the stately slowness of duration, the moving wonder of every hour," is the way Impressionism and its phenomenologists put it, Claude Monet among them. In so doing, between 1899 and 1901, they prefigured the theory of relativity, Albert Einstein's theory about point of view. Monet, the man who painted the celebrated "Series" even decreed that: EVERYTHING CHANGES, EVEN STONE, thereby forecasting the emergence of kinematic energy alongside kinetic energy. The Lumière Brothers would develop kinematic energy more fully, thereby anticipating the mutations, in the twentieth century, of the multifractal spacetime of which Mandelbrot—that other impressionist!— would become the secular prophet.[96]

Drómos, *phusis*, *kīnēsis*, and *phōs* feature prominently in this passage. The Lumière Brothers—*lumière* translates *light* or *lighting*—were pioneers in film and cinematography. Hence, The Light Brothers forecasted the emergence of modern cinematic—*kinematic* and *kinetic*—energy. What is especially important to note is how Monet, and impressionism, represent Virilio's positive account of approaches to speed, physics, motion, and light. In contrast to the other figures mentioned here—the cinematographic Lumière Brothers, the relativity physicist Albert Einstein, and the fractal geometrician Benoît Mandelbrot—the impressionists represent Virilio's ideal form and posture for ethics. It is their attentiveness to the beauty and power of speed, physics, motion, and light, which Virilio admires and uses for his theology, ethics, and aesthetics. So, related to the question of *how* one is redeemed by Jesus Christ and can live this impressionist way of life, Virilio is clear at other points in his writing. For example, in *The Vision Machine*, he writes:

> We have all had enough of hearing about the death of God, of man, of art and so on since the nineteenth century. What in fact happened was simply the progressive disintegration of a faith in perception founded in the Middle Ages, after animism, on the basis of the unicity of divine creation, the absolute intimacy between the universe and the God-man of Augustinian Christianity, a

> material world which loved itself and contemplated itself in its one God. In the West, the death of God and the death of art are indissociable; *the zero degree of representation* merely fulfilled the prophecy voiced a thousand years earlier by Nicephorus, Patriarch of Constantinople, during the quarrel with the iconoclasts: "If we remove the image, not only Christ but the whole universe disappears."[97]

Reception of the Image of God, Jesus Christ, redeems. The impressionist artists formally model what this reception and redemption looks like. For this reason, the impressionist artist at rest in the environment models for Virilio, and for the believer, what it looks, sounds, feels, smells, and tastes like to be at rest in creation and to have one's senses thoroughly mediated by Jesus Christ: removing oneself from modernity's race (*drómos*), ceasing agitated movement (*kīnēsis*), standing at a distance from false light (*phōs*), and being filled with God's light (*phōs*) in a creation (*phusis*), which speaks to the people who can sense creation's speech through the mediation of Jesus Christ; the one whom through all things came into being, Word become Flesh, Light shining in the darkness, as revealed in John 1:1–4. In short, Virilio's impressionist way of life is best described as a Johannine impressionist way of life; an impression of God, others, and creation mediated by Jesus Christ.

Virilio does not elucidate a doctrine of mediation but the way in which he conveys his understanding of mediation certainly is valuable for theology, ethics, and aesthetics. Furthermore, given his frequent references to Bonhoeffer's *Creation and Fall* and *Discipleship*, answers to questions about Virilio's theology of mediation may be found in Bonhoeffer's work. My suggestion is that Bonhoeffer gives words to the grammar of Virilio's Christ-mediated impressionism. For instance, in his *Ethics*, Bonhoeffer writes, "*Jesus Christ, the eternal Son with the Father in eternity*—this means that nothing created can be conceived and essential understood in its nature apart from Christ, the mediator of creation. Everything has been created through Christ and toward Christ, and everything has its existence only in Christ (Colossians 1:15ff.) Seeking to understand God's will with creation apart from Christ is futile."[98] As the previous section demonstrated, the attention Virilio gives to Bonhoeffer's "sicut deus" descriptions implies that he also appreciated, and implicitly worked with, Bonhoeffer's "imago dei" and "agnus dei" descriptions.[99] Taken together, it is not a stretch to say that Bonhoeffer's account of mediation significantly informed Virilio's. Not only does an appreciation of Bonhoeffer help fill out Virilio's theology of mediation, Bonhoeffer also constructively mediates, as it were, between Ellul's emphases on hearing and Virilio's emphases on senses. Virilio uses the Lutheran Bonhoeffer, it seems, to mediate between his Catholicism and Ellul's Protestantism, which was heavily influenced by the Calvinist Barth.

In other texts, Virilio describes the form this Johannine impressionist way of life can assume. For example, Virilio says, "Everything is in looking and in the word, but through the body's senses in their entirety. Now, the Bible is the Word. You have to read the prophets and read [Jesus] Christ —'live' them through your senses and your intelligence."[100] Again, this time in positive terms, Virilio forms connections between theology, ethics, and aesthetics, between Word and Flesh, and between sound and sense, because Jesus Christ's mediation brings Word, Flesh, sound, sense, theology, ethics, and aesthetics together. The Johannine impressionist way of life consequently forms an image of what it looks, sounds, feels, smells, and tastes like when Word and Flesh, sound and sense, are unified and embody a morality unwanted by the world, but given to a scattered and small in number people. First, the scattered and small in number people of Israel are assembled and led by God. Then, the church, drawn together, impressed upon, by Jesus Christ, Word become Flesh, Light in the darkness.

Not only is there the rest of the impressionist way of life. There is also the silence. Virilio says the following about the silence:

> Silence is a voice. I'm obviously paraphrasing [André] Malraux, but mainly the Old Testament, especially Psalm 18: language without speech. God speaks without speech before the Prophets. Before Israel. God speaks in Creation: the sun speaks to the night, the night to the day. Clearly, the language without speech is the language of Creation. And Creation is nature. It's the sun, the beauty of the sea, etc., but it is also the creation of humanity. It's the silence of painting, the silence of the Impressionists. The silence was thwarted, and definitively in my opinion, by the arrival of the talkie—not by the arrival of cinema, but by the *talking* cinema.[101]

Virilio's ethics of Christologically mediated impressionism requires a seeking out of silence and rest in the world, which tends to move in the opposite direction—toward constant, frenetic, speed (*drómos*), motion (*kīnēsis*), and light (*phōs*) in every possible environment (*phusis*). By being moved by Jesus Christ toward silence and rest, one can relearn, or perhaps learn for the first time, how to cultivate the practices of silence and rest. This can transform for the believer and the church the seemingly trivial into the productive, cause the silent to speak, and cause the closed to open. As the passage just cited shows, as well as Virilio's writing as a whole, Virilio is just as wary of the mediations of technology as Ellul is. However, Virilio's wariness of the mediations of technologies does not cause him to evacuate mediation from his theology, ethics, and aesthetics as much as Ellul does. He is therefore an important resource in suggesting how Ellul's emphases on a hearing of and response to God's Word might take root in the creation in which creatures always are set.

Taken together, the impressionist way of life's rest and silence starts with Sabbath, a day dedicated to removing oneself from modernity's race (*drómos*), ceasing agitated movement (*kīnēsis*), standing at a distance from false light (*phōs*), and being filled with God's light (*phōs*) in a creation (*phusis*), which speaks to the people who can sense creation's speech through the mediation of Jesus Christ. With these emphases in view, Virilio writes in *Lost Dimension*:

> Continuous time is that of chronology or history, it certainly is not that of everyday experience. Interruptions of activity or productivity are essential to the structuring of real time, for individuals and for social groups, and here the day is the reference point and the primary standard for these structuring interruptions, as is evidenced in common expressions such as "to see the light of day" or "to call it a day." More than any other physical, urban, or architectonic barrier, and more than any other natural or political border, the day marks off difference of temporality, differences of regimes that affect the consciousness of time passing, not only though rhythms of sleeping and waking but more with the eternal return of daylight and of night. Thus, the region and the city are not organized exclusively through a cadastral system of blocks, neighbourhoods, city centers and peripheries, or clusters apartment complexes. They are also arranged through a calendar system of vacations, leaves-of-absence, and holidays off. Some interruptions return us to the religious purposes of the Sabbath or the Sunday rest, the seventh day of Creation, a vacation dedicated to contemplation, to the reception of the finished product.[102]

The impressionist way of life is initiated by and cultivated on Sabbath, the Sunday rest, the seventh day of Creation, a reception of God's finished product, and with that reception, a resistance to other regimes of the world, which affect the consciousness of time passing. The vacation of Sabbath rest and silence is not, therefore, a vacation as the so-called vacation industry would have it. Rather, the Sabbath vacation is a vacating, an emptying, of frenetic movement and constant speech, a space and time filled by an opening up to receive rest and silence as a gift from God; God's gift of the ability to rest and to be silent in a world that cannot seem to rest and cannot seem to stop talking. Believers and churches who refuse to initiate or cultivate Sabbath rest and silence, then, will be less likely to interrupt and resist the other regimes, the regimes of technology, which affect the consciousness of time passing, and as a consequence, will be less likely to live and to cultivate the Johannine impressionist way of life.[103] From this, the true, Jesus Christ mediated, guidance for thought (*epistēmē*), action (*poiēsis*), and making (*tékhnē*) in creation can spring up from the ground.

RACING TOWARD A BUNKER

Virilio's lifelong study of the architectural form of the bunker and his descriptions of the inspiration for the design of *The Bunker Church*, when combined with the theology and ethics examined in the previous sections, offer theological and ethical resources for Christians and churches to transform even the most horrible habitats of war, terror, and death into habitats of peace, rest, and life.

Virilio's fascination with and fear of the bunker began in his childhood and persisted throughout his life and writing. In *Bunker Archaeology*, he writes the following about the thirty years he spent exploring the Atlantic coastlines to study and photograph The Atlantic Wall, some 1,500 bunkers built by Germany's Third Reich and abandoned after World War II:

> The space I was charting with surveys and measurements of different types of casemates was the space of a different historical time than that of the moment of my trip; the conflict I perceived between the summer of seaside bathing and the summer of combat would never again cease. For me the organization of space would now go hand in hand with the manifestations of time. This actual archaeological break led me to a reconsideration of the problem of architectural archetypes: the crypt, the ark, the nave.... The problems of structural economy had become secondary, and now I would investigate the Fortress Europe, which was vacant from now on, with an eye to the essence of architectural reality.[104]

Later, he infuses these studies of the bunker with resurrection imagery:

> Contemplating the half-buried mass of a bunker, with its clogged ventilators and the narrow slit for the observer, is like contemplating a mirror, the reflection of our own power over death, the power of our mode of destruction, of the industry of war. The function of this very special structure is to assure survival, to be a shelter for man in a critical period, the place where he buries himself to subsist. If it thus belongs to the crypt that prefigures the resurrection, the bunker belongs too to the ark that saves, to the vehicle that puts one out of danger by crossing over mortal hazards. Literally, casemate means "strong house," reinforced house; it is always a case of habitat, or rather of a kind of clothing, of collective armor in the final analysis.[105]

Resonating with Ellul's descriptions of technology as a protective envelope, Virilio's phenomenological descriptions of the bunker's form reveal the degree to which modernity, with its total wars and powerful technologies, blurs architectural and archetypal forms, which Virilio defines here as organizations of space interwoven with manifestations of time. Virilio urges readers to ask: *Is the bunker really a bunker? Or is it a tomb, an ark, or a house? What will be waiting whenever whatever it is, is reached?* In light of what the previous chapters have shown about Virilio's writings on technology, and based on his phenomenological descriptions of the bunker's form in these passages, the answer given to this question depends on the light (*phōs*) in which the bunker, or any architectural form, is perceived. Here, Virilio writes that the "crypt [prefigures] the resurrection" and the "bunker belongs [to] the ark that saves." So, in the light (*phōs*) of the resurrected Jesus Christ, the dark crypt of death is filled with the dazzling light of life; the bunker being barraged by shells assumes the form of the ark that saves from the deluge. Another devastating image turned to hope affected by Virilio. This time, the turn is concretized in the form of the bunker, which Virilio says, "Is the reverse figure of the destructive power of the twentieth century. Auschwitz, Hiroshima, both." And "that's why the bunker, to me, is the symbol of modern times."[106] All of these passages give glimpses of the theological, ethical, and aesthetic reasons for Virilio's design for *The Bunker Church* in Nevers, France. A bit more context will reveal how Virilio generated such a positive theological, ethical, and aesthetic turn out of such a negative image. First, he describes his negative encounters with bunkers as a child during World War II:

> To avoid using electricity—you know that it will most likely be interrupted during a bombing—they painted the walls of the [air-raid] shelters with phosphorous. At first people flocked to them by the hundreds, and in the end stopped going, preferring to die in the streets. Why? Because they were on top of one another, as in a subway. And now and again the fans would stop on account of the bombs falling. People would suffocate. And with this phosphorous light you literally had an image of Dante's *Inferno*. To me the bunker is total war.[107]

At the same time, Virilio found hope in these hellish bunkers of horror and destruction. In the same interview, he describes in positive terms the inspiration for *The Bunker Church*:

> In Dusseldorf [after the war] some neighborhoods still were in ruins, and an [air-raid shelter] had been turned into a church. I went to mass in a bunker that is called the Church of the Holy Sacrament. Seeing a place like that Christianized, a place of terror, haunted by fear, that's what really interested me. And so when I came back [to France], I realized that in reality, nuclear terror had only just begun. Those were the days of *Atomic Cafe*. Everyone was building bomb shelters. And I decided that the grotto at Lourdes was today's bomb shelter. It is the place of horrors, the place of great fear, the end of the world. So I drew inspiration from the bunker to do the job. I chose the shape of a heart, the double ventricles, split in two, cut down the middle, broken. One of them is the choir for communion, and the other the choir for confession.[108]

Virilio's ethics of impressionism, combined with his theology of mediation, enabled him to sense a transformation of bunker's aesthetic form, a habitat of war, horror, and death, into the theological, ethical, and aesthetic form of a church, a habitat of peace, rest, and life, at the heart of which the Word of God is preached, believers confess and repent of sin, and the Body and Blood of Jesus Christ is received in the form of the Eucharist. In this way, *The Bunker Church* is a theological, ethical, and aesthetic concretization of Virilio's understanding of the forms thought (*epistēmē*), action (*poiēsis*), and making (*tékhnē*) ought to assume for the individual Christian and the institutional church today. To what are people and things—to what are *we*—racing, in the end? A tomb, a temple, an ark, a church, or a bunker? Perhaps a combination of each, at once, in the same time and space. Ultimately, it depends on the light in which the habitat, the organization of space confronted by the world's manifestations of time, is sensed, and how believers and the church concretize that sense theologically, ethically, and aesthetically in the present. But, it must be said that for the believer and for the church, the race necessarily starts with a rolled-away stone; a flash like lighting; with wind sprints toward an empty tomb.[109]

Virilio does not believe all churches ought to appear in the form of a heart-shaped bunker. In another interview, he remarks, "One must remember that [*The Bunker Church*] was dedicated to St. Bernadette of Lourdes. Naturally, had it been dedicated to St. Francis of Assisi, it would have been a different church. It would have been a transparent church, a church for the birds!"[110] The church's aesthetic form can change depending on the time and space, but the Word preached, the confessions made, and the Eucharist received remain consistent, because it is from their unified consistency that contingent aesthetics and ethics spring. Describing the Eucharist in one of his earliest essays, Virilio writes:

> Christ is twice host. The day before his death, he defines the object of sacrifice by taking and distributing bread and wine: this is me, what? the ordinary, the everyday, I am the principle of what is to all, their freedom which is banality, the most common is the most divine. The next day, the second host shows that the daily life which is Christ is incompatible in its principle of common liberty with an order which, by its conservative and prophylactic construction, has withdrawn from human time and space, from the everyday, [and] it can be thought today, in all likelihood, *until the consumption of centuries or their consummation.* At Emmaus, conversely, Christ re-emerges from the shadow of banality, the blindness of the disciples ceases when he renews the act [*le geste*] of *transgression*, this gesture [*ce geste*] which the Church covered with gold, mud, and innocent blood nevertheless has repeated over the centuries, distorts, adulterated, and yet, still faces the State of the West as the sign become almost unconscious and mechanistic of ultimate resistance to the inventor of the colony, the ghetto, the camps, the reserves, the proletarianization, the underdevelopment, even worse,

> the destroyer of sociality, the socialism of poverty, its products, its inventions, its cultures, its assets, [and] its techniques.[111]

Jesus Christ's act/gesture of transgression, the rending of the bread/body and the flowing of the wine/blood, gives freedom, transforms the most common into the most divine, cures blindness, and, in spite of failures of the church throughout history, is a sign of resistance to all that is wrong in creation crying out to God as it awaits the consummation of the Eschaton. This, for Virilio, is what makes the silent speak and the closed open. This, for Virilio, is what makes the impressionist way of life what it ought to be. This, for Virilio, is what theologically thickens, in a Eucharistic sense, his constant examination of the relationship between the accident and the substance. The unification of accident and substance in the bread/body and blood/wine illumines the integral relationships between all accidents and all substances in creation. The wounds of war's projectiles are stopped and healed by the bread and the wine implanting peace in the Church body, which is Jesus Christ's.[112] In this way, from the effectively leveled ground of Ellul's iconoclasm, desacralization, and demythologization, inspired by a theology of justification by faith in Jesus Christ alone, flowing into an ethic of being clothed and fed by God, Virilio builds an ethics and aesthetics of Johannine impressionism formed and filled by a theology of Jesus Christ's mediation.

The next sections add to, and at certain points move beyond, the ethics examined and developed thus far. Ellul and Virilio obviously offer vital resources to theology and ethics today, but at the same time, their ethics for the *ekklesia* are nascent if not altogether thin. But this is not a consequence of either individual's lack of involvement in the church. Both Ellul and Virilio actively served in their ecclesial communities. Rather, as it has been shown thus far, the absence of a detailed account of the *ekklesia* in both scholars' work is a consequence of their socio-historical situation. As they thought and wrote after their encounters with the horrors of the Holocaust and Hiroshima, their suspicion of large gatherings of people and the influence of social institutions focused their critical attention upon *the individual.* Accordingly, Ellul and Virilio wanted to inspire and equip *individuals* to resist the mass manipulation and violence that they witnessed before, during, and after the war. Now that the voices of Ellul and Virilio have been heard, the next sections move their more individually focused ethics toward a more ecclesially focused ethics after considering the value of both authors' writings for today.

TODAY'S PROBLEMS, THREATS, AND QUESTIONS

Both Ellul and Virilio focus on the individual in their writings to the extent that it remains to be seen how their ethics might be articulated ecclesially.

While both scholars had good reasons to focus on the individual instead of the group in their work, it is also important to develop ethics for the believers who gather together to make up the church. To that end, this section uses themes examined throughout this book to develop an ecclesial ethic for a world filled with powerful, idolized technologies.

In our hyper-technologized world, Ellul and Virilio are ignored at our peril. Their rigorous, skeptical, and thoughtful scrutiny resists the zeitgeist and idolatry, which characterizes modernity in ways that evoke the prophetic work of theology and ethics. In fact, their proleptic and prophetic insights into how technologies alter thought, making, and action in the world have been validated ever since their initial publication. With this in view, I want to consider the importance of these scholars' writings on technologies for theology and ethics today. Drawing on 1 Corinthians 12:12–28, the more individually focused ethics of Ellul and Virilio will be reoriented toward an ethics more anchored in the *ekklesia.*

As far as churches are concerned, I think the current problems, threats, and questions that appear with any sincere consideration of Ellul and Virilio's work on technology are tangled around the issue of power and the forms of idolatry manifested by humanity's enduring lusts for power. Putting this problem and threat in the form of a question, Ellul and Virilio urge those of us who consider themselves a part of the people of God: how ought individual Christians and institutional churches seriously think about and intentionally respond to the power and idolatry of technology that these two thinkers expose with their sociological, historical, philosophical, phenomenological, and theological analyses? The voices of Ellul and Virilio cry in response to this question from postwar France. Historians have heard these cries, whose continued relevance in the present realities of a globalized world shaped by powerful technologies means that ordinary Christians ought to begin listening intently. Their arguments arrive to us as if forged for just our moment. *Propaganda*, both historically and presently, is a bigger part of everyday life than most people—including academics, Christians, and churches—would prefer to perceive or admit. In *The Cultural Cold War*, Frances Stonor Saunders writes:

> Official documents relating to the cultural Cold War [waged by the CIA] systematically undermine [the] myth of altruism. The individuals and institutions subsidized by the CIA were expected to perform as a part of a broad campaign of persuasion, of a propaganda war in which "propaganda" was defined "as any organized effort or movement to disseminate information or a particular doctrine by means of news, special arguments, or appeals designed to influence the thoughts and actions of any given group." A vital constituent of this effort was "psychological warfare," which was defined as "The planned use by

a nation of propaganda and activities other than combat which communicate ideas and information intended to influence the opinions, attitudes, emotions, and behavior of foreign groups in ways that will support the achievement of national aims." Further, the "most effective kind of propaganda" was defined as the kind where "*the subject moves in the direction you desire for reasons which he believes to be his own.*" It is useless to dispute these definitions. They are littered across government documents, the [data] of American post-war cultural diplomacy.[113]

Further conforming Ellul's analyses of propaganda, particularly his argument that the intellectuals and artists of society are most vulnerable to the influence of and *myths* manipulated by *propaganda*, Saunders writes, "The defense mounted by the custodians of the period—which rests on the claim that the CIA's substantial financial investment came with no strings attached—has yet to be seriously challenged. Amongst intellectual circles in America and western Europe there persists a readiness to accept as true that the CIA was merely interested in extending the possibilities for free and democratic cultural expression."[114] Compare this to what Ellul wrote in 1967 in *Propaganda*: "Naturally, the educated man does not *believe* in propaganda; he shrugs and is convinced that propaganda has no effect on him. This is, in fact, one of his greatest weaknesses, and propagandists are well aware that in order to reach someone, one must first convince them that propaganda is ineffectual and not very clever."[115] Ellul saw in advance and with acuity what the documents of history retrospectively and conclusively are revealing today: the most propagandized people tend to be the ones who do not perceive themselves as propagandized; the most idolatrous people tend to be the ones who do think themselves idolaters; and, that intellectuals are the ones who are most prone to propagandization and idolatry. The challenge today, then, is for Christians, churches, and those who are part of the theological academy, which serves the church, to do what Ellul and Virilio did in their time: expose *propaganda* and resist it in thought, writing, and action; to demythologize the *myths*, which institutes the sacred orders of society that should never be sacralized; and to level to the ground any idolatrous justification other than the justification of Jesus Christ. Previous sections in this chapter have derived ethics from Ellul and Virilio's writings to help meet his challenge, but the critical question of *how* remains: How to expose and resist *propaganda* today; how to demythologize the *myths* instituting the sacred of orders in society today; how to level all idolatrous justifications other than the justification of Jesus Christ today?

My hope it that this book will show readers that answering these questions from Ellul and Virilio's work turns on how both thinkers are read. Reading Ellul and Virilio as thinkers in engagement with, rather than aloof from, each other and theological tradition is crucial because it reveals valuable themes and resources in their work, which have not received adequate attention. Of course, Ellul and Virilio's work with and reworking of theological tradition is far from traditional (i.e., typically theological). However, the unique, at times unruly and quixotic character of their work, when combined with their disciplines of choice, model how to expose and resist *propaganda* today; model how to demythologize *myths* today; and, model how to level idolatrous justifications and mediations other than the justification and mediation of Jesus Christ today. Beyond modeling this, Ellul and Virilio show us where to begin this work: returns to zero. Chapters 3 and 4 especially showed how "the origins which [modern] myths are telling us about is no longer the same origin, nor is the event which they are interpreting the same."[116] And in light of this, Ellul and Virilio teach us to ask and answer: What are the origins modern *myth* are telling us about today, what sacred orders are those *myths* instituting, and how ought we to respond theologically and ethically?

In Ellul's and Virilio's cases, their nascent theological tradition engaged their broader theological tradition, and reworked it as they identified and responded to the problems, questions, and powers of their time. In other words, this nascent theological tradition *performs* for us how theology and ethics might perform a similar engagement with and reframing of theological tradition today, in light of the problems, questions, and powers of our time, with their specific, unique, contributions to theology and ethics being the interdisciplinary war they individually and collectively waged on idolatry in the modern world. So, what are the winds of deceitful doctrine tossing to and fro ordinary Christians today, and how might attentiveness to social *myth* today mark the concentration points for the winds of sacralized, deceitful, doctrines?[117] How might theology and ethics respond? Ellul's structural method for analyzing society, combined with his understanding of the fully foreign force of God's revelation, offers a good place to start when answering these questions. Virilio saw this in Ellul's writing, and he similarly started from zero, taking critical cues from Ellul.

The connections between commonplace technologies, violence, and war are always tighter than most people, perhaps especially academics, Christians, and churches would prefer to perceive or admit. In *Imagineers of War*, Sharon Weinberger notes:

> By the 1990s [popular] news accounts were frequently repeating as accepted truth that the Internet's origins could be traced back to [the Advanced Research Project Agency's] work on creating a military communications system that could survive nuclear war. This account sparked a counter-narrative by ARPA-funded scientists who insisted that computer networking was pursued primarily for its civilian applications. The truth is more complicated, and it is impossible

> to divorce the Internet's origins from the Pentagon's interest in the early 1960s in the problems of war, both limited and nuclear. The Internet would likely not have been born without the military's need to wage war, or at least it would not have been born at ARPA.[118]

Compare this to what Virilio wrote in *The Information Bomb*, published in 1998:

> Has not the network of networks [the Internet] established on the basis of the ARPANET system, which was designed to resist the electromagnetic effects of an atomic war, presented an occasion—since the Gulf War—for launching the very first *universal advertising campaign* for a systemic product which interests *no one in particular* and *everyone in general*? The promotion of the Web and its on-line services, an unprecedented phenomenon of ideological contamination, no longer bears any resemblance to the marketing of a practical technology, such as the sale of a vehicle or even of some kind of broadcasting equipment (radio, television), since what is involved in this case is the *most immense enterprise of opinion transformation* ever attempted in "peacetime," an undertaking which has scant regard for the collective intelligence or the culture or nations.[119]

Again, the challenge for Christians and churches today is to do what Virilio and Ellul did in their time: expose violence and resist it in thought, writing, and action with a Christ-justified-and-mediated way of life. Not all *technological* discourse is *propaganda*, not all uses of technology are violent, and not all technologies ought to be rejected. There certainly are aspects of the writings of Ellul and Virilio, some of which have been featured in this book, which imply they do think technology ought to be rejected wholesale. But as the open chapter showed, their experiences before, during, and after World War II are what made their writings about technology so emphatic. Which is to say, because of their encounters with the power and idolatry of technology, they wrote about technologies to shock us and to provoke critical shifts in our understanding of what we think we know, and do not know, about technology.[120] For this reason, both scholars' warnings about the insidious effects of technology ought to be taken seriously because they collectively urge readers to admit that all people, including Christians, churches, and theologians are always closer to extreme degrees of manipulation, idolatry, and violence than they might initially think or sense—until the serious consequences begin to appear.

The argument presented here has sought to highlight the extent to which Ellul and Virilio expose how it is the utility and power of technologies, at base, that offers the most dangerous temptations for Christians—both individually and communally. After World War II, there were significant increases in the number of North American Christians and churches optimistic about and willing to use the power of technology to proclaim the gospel. In most instances, these efforts were motivated by the best of intentions, but there were situations where the motive—explicit or buried—was to claim influence, authority, and wealth. Such phenomena, I hope to have shown throughout this book, cultivate fertile soil for idolatry and violent power. And in fact, now, in the twenty-first century, the long-term consequences of these best and worst cases and intentions are appearing and becoming quite similar, in fact, to those of twentieth-century Europe. Nations are antagonizing nations, nations are internally divided, institutions increasingly are coopted or crippled by state power, and disillusionment, rage, and violence are fomenting between groups and individuals.[121] The difference is that today all of this is happening at a simultaneously global and local level, powered by rapidly developing technologies, with Christians and churches tossed to and fro, often becoming complicit, in the midst of it all.

This context suggests it is wise to take up again Jesus's question, as did Ellul and Virilio did in their time: What will it profit a person or an institution if they gain the whole world and forfeit their life?[122] To seize, use, and worship without restraint, the power technology offers to harmfully gain anyone or anything, including people, authority, and wealth, always risks forfeiture of the life of the believer and the life of the church. Jesus Christ, after all, told the crowds and the disciples, just before asking this question: "If any want to become my followers, let them deny themselves and take up their cross and follow me. For those who want to save their life will lose it, and those who lose their life for my sake will find it."[123] Such self-denial and taking up of a cross may be difficult when tempted by easy paths to power, but it is unavoidable. Self-denial and taking up of a cross is made possible by Jesus Christ. It leads to life, faith, hope, and love, rather than death, despair, nihilism, and rage. Manipulative and destructive power is not desired, needed, nor feared when believers truly live in the light, salvation, and strength of God, who is Father, Son, and Spirit. This is the truth to which writers like Jacques Ellul and Paul Virilio witness and this is why their writings are vital for theology and ethics today.

FROM INDIVIDUAL BODIES TO ECCLESIAL BODIES

There have been hints at how the ethics of Ellul and Virilio might move from a focus on the individual mind and body to a focus on the ecclesial mind and body, but there is not yet a full account in this book of how such an ethics could be formed. With 1 Corinthians 12:12–28 as its scriptural foundation, this section moves these scholars' more individually focused ethics to a more ecclesially focused ethics.

As mentioned in the previous section, technology has put the world in a precarious place, and Christians and churches, whether they perceive it or not, are in and bear a particular responsibility to this world. Ellul and Virilio offer practical resources that can help communities navigate the theological, ethical, and aesthetic path before them. Both thinkers reframed theological tradition to respond faithfully to the world in which they found themselves. This, I would argue, models a way for theology and ethics to enact similar reframing of theological and ethical tradition today. Ellul and Virilio offer thick, dynamic, and applicable resources for individual Christians and institutional churches aspiring to think about and use technologies in ways that line up with, rather than diverge from, the mission to which God has called them.

How then, might an ethics of being justified, clothed, and fed by God, combined with an ethics of Christ-mediated impressionism, take shape in ecclesial bodies today? As suggested earlier, the writings of Ellul and Virilio are more nascent than comprehensive in respect to ecclesiology. This is not because they eschewed service in and outside the church and not because they did not see a value in ecclesiology. Rather, it is because both thinkers spent much of their time writing to and for as well as directly serving disenfranchised individuals both in and outside the church body. As a consequence, their insights rarely were welcomed by the power brokers of their society. To render a service to their work, the remainder of this section will aim, with 1 Corinthians 12:12–28 as its scriptural foundation, to move their more individually focused ethics to a more ecclesially focused ethics. Beginning with verses 12–13:

> For just as the body is one and has many members, and all the members of the body, though many, are one body, so it is with Christ. For in the one Spirit we were all baptised into one body—Jews or Greeks, slaves or free—and we were all made to drink of one Spirit.

Everything that has been said about the ethics Ellul and Virilio form for the individual body can and must be said for the ecclesial Body of Christ fastened together by the Holy Spirit. Stated negatively, this means whenever or wherever technology is used to manipulate an individual member of the ecclesial body, even if it is just one member, the entire ecclesial body is manipulated. This, in turn, determines how the ecclesial body does and does not reach out to and embrace people in the world who are not yet part of the ecclesial body. Stated positively, when an individual member is justified, clothed, and fed by God, and lives a Christ-mediated impressionism, the entire body shares in that justification, clothing, feeding, mediation, and impressionism. Again, this, in turn, determines how the ecclesial body does and does not reach out to and embrace people in the world who are not yet part of the ecclesial body.

Precisely how all of this happens in and outside the ecclesial body, both in negative and positive terms, is a mystery. But, it is true, individually and ecclesially, nonetheless. To consciously reject or unconsciously dismiss this mysterious truth is to risk the consequences, which come with any manipulations of the ecclesial Body of Christ and those who are outside of this body. On the other hand, to rejoice in this mysterious truth is to share in the joy of being justified, clothed, fed, mediated, and impressed as a member of the ecclesial Body of Christ fastened together by the Spirit, and in turn, to see others who are not yet members of this body become members of this body. In sum, the ecclesial body is a body, and for this body to remain the Body of Christ fastened together by the Spirit, it ought not to become a cyborg controlled by manipulative technologies that enhance some members and replace others.[124] Now, verses 14–22:

> Indeed, the body does not consist of one member but of many. If the foot would say, "Because I am not an eye, I do not belong to the body," that would not make it any less a part of the body. If the whole body were an eye, where would the hearing be? If the whole body were hearing, where would the sense of smell be? But as it is, God arranged the members in the body, each one of them, as he chose. If all were a single member, where would the body be? As it is, there are many members, yet one body. The eye cannot say to the hand, "I have no need of you," nor again the head to the feet, "I have no need of you."

As Ellul and Virilio commit to either justification or mediation, they tend to focus on hearing or seeing and the mind or the senses in their writings on technologies. These verses are a reminder that there is a diversity of members in the one ecclesial body and that this pluriformity should be celebrated and cultivated rather than discouraged and dissolved. Some members will and should, like Ellul, be the ear rather than the eye, the hand, or the foot. Other members will and should, like Virilio, be the eye rather than the ear, the hand, or the foot. Other members will and should be the hands and the feet rather than the ear or the eye, and so on. To elevate one member over another, or to act implicitly or explicitly as if one member is needed while another is not, consequently is to risk sequestering or severing members from the one ecclesial body.

A particular risk with technologies is, as Ellul and Virilio collectively and compellingly demonstrate, that technologies tend to favor and amplify certain members—usually the ears, the eyes, the mouths, and the hands—to the exclusion of all others. This, in turn, is a threat to all members. It is by hearing and seeing what both Ellul and Virilio have to say and show in their writing, then, that each member in the church can be seen, heard, celebrated, and strengthened in his or her being a particular member within the one ecclesial body. It will help the ecclesial body remain the Body of Christ fastened together by Spirit, and also, help ecclesial bodies resist inclinations to become cyborgs controlled by manipulative technologies, which enhance some members and replace others. Finally, verses 22–27:

> On the contrary, the members of the body that seem to be weaker are indispensable, and those members of the body that we think less honorable we clothe with greater honor, and our less respectable members are treated with greater respect; whereas our more respectable members do not need this. But God has so arranged the body, giving the greater honor to the inferior member, that there may be no dissension within the body, but the members may have the same care for one another. If one member suffers, all suffer together with it; if one member is honored, all rejoice together with it. Now you are the body of Christ and individual members of it.

This is where fallen conceptions of honor and dishonor, superiority and inferiority, strength and weakness are inverted in and by the ecclesial Body of Christ fastened together by the Spirit. As the previous paragraphs noted, a particular risk with technologies that Ellul and Virilio expose is that technologies tend to favor and amplify certain members—usually the ears, the eyes, the mouths, and the hands—to the exclusion of other members. The added risk with technologies in light of these verses is that technologies also tend to favor and amplify who and what the fallen world elevates as honorable, superior, and strong, while excluding and devaluing, if not crushing altogether, who and what the fallen world condemns as dishonorable, inferior, and weak. In contrast, the Body of Christ, fastened together by the Spirit, with no internal dissension, ought to live in and out the truth that, as Paul writes in 1 Corinthians 1:27–29, "God chose what is foolish in the world to shame the wise; God chose what is weak in the world to shame the strong; God chose what is low and despised in the world, things that are not, to reduce to nothing things that are, so that no one might boast in the presence of God." Indeed, it often is who and what the fallen world condemns as dishonorable, inferior, and weak, both in and outside of the ecclesial Body of Christ fastened together by the Spirit, that reveals to the world what it really and truly looks like to be justified, clothed, and fed by God, and live a Christ-mediated impressionism. It is, after all, Christ crucified that is a stumbling block and foolishness for some, but the power and the wisdom of God for others.[125] What would it look, sound, smell, taste, and feel like, then, for ecclesial bodies to apply these truths to their understandings and uses—or non-uses—of technologies today? Such sights, sounds, smells, tastes, and touches only appear when fallen, manipulative, power is rejected, and the otherworldly, power—the grace—of Jesus Christ is received within and witnessed to freely, both in and outside the one ecclesial body, in faith, hope, and a love.

We are not our own, we were bought at a price. With bodies that are temples of the Holy Spirit, who is in us, we therefore honor God with our bodies.[126]

The inversion of worldly honor and dishonor, superiority and inferiority, strength and weakness, begins practically with ecclesial honoring, uplifting, and strengthening of the members and the organs who are dishonored, excluded, and weakened by the world. Only a seeing and a hearing of the presence and perspective of the powerless can challenge and change poor perceptions and handlings of power. Jesus must be taken seriously when he says, "Truly I tell you, just as you did it to one of the least of these who are members of my family, you did it to me" (Matthew 25:40). The call of the hungry, the thirsty, the stranger, the naked, the sick, and the prisoner all demand ecclesial response: listening, honor, encouragement, and strengthening.[127] The young and the differently abled also.[128] Anyone excluded from having their presence seen and their voice heard in discussions of power must be included, have their presence seen, and have their voices heard.[129] This is where such ecclesial inversions of worldly power start today, and this is where the identification, exposure, and dismantling of idols in our world starts today. Jacques Ellul and Paul Virilio's nascent theological tradition helped me arrive at this conclusion, just as theological tradition as a whole has. For, what theological tradition within the broad and deep Christian tradition is not nascent in its own way? If this book has represented Ellul and Virilio's work on theology, ethics, and technology well, it will have shown that all theological traditions are nascent and maturing as they anticipate the return of the one to whom all of our theological traditions direct witness and worship: Jesus Christ, the Word of God in the Flesh.

NOTES

1. Ellul, *Théeologie et Technique*, 157.
2. Ellul, *Technological Society*, 24–25. Virilio uses magic imagery in similar ways, though his method and focus differs. See, for example, Virilio, *Speed and Politics*, 112; Virilio, *University of Disaster*, 89–90; Virilio, *Administration of Fear*, 82; and, Virilio, *The Great Accelerator*, 25–26.
3. André Leroi-Gourhan, *Evolution et techniques, vol. 2: Milieu et techniques* (Paris: Albin Michel, 1945).
4. Ibid., 198 and 241.
5. Ibid., 384.
6. Genesis 3:20–21 reads, "The man named his wife Eve, because she was the mother of all living. And the LORD God made garments of skins for the man and for his wife, and clothed them."
7. Ellul, *To Will and to Do*, 11 for a discussion of postlapsarian nakedness and shame. Ellul references Dietrich Bonhoeffer, *Ethics*, ed. Clifford J. Green, trans.

Reinhard Krauss, Charles C. West, and Douglas W. Stott (Minneapolis: Fortress Press, 2009), 214–217 and 302–307.

8. Ellul, *To Will and to Do*, 195. Further discussion of necessary work and works after the Fall appears in *Reason for Being*, 93–105.

9. Colossians 3:5–17 elucidates the ethics stemming from this theology. See also Luke 24:49; Ephesians 4:24; and 2 Corinthians 5:1–4.

10. See chapter four for further discussion on Ellul's reading of the Fall in *Humiliation of the Word*.

11. Ellul, *Humiliation of the Word*, 94–95. Also, Ellul, *Meaning of the City*, 99; Ellul, *Living Faith*, 257–158; and Ellul, *Freedom, Love, and Power*, 77–80.

12. Ibid.

13. Ibid. See also, Ellul, *To Will and to Do*, 11–13 and Ellul, *Théologie et Technique*, 152–153.

14. Ellul, *Ethics of Freedom*, 54.

15. For biblical passages that affirm these ethics, see Luke 24:49; Romans 13:14; Ephesians 4:24; Colossians 3:10–12; Galatians 3:27; and 2 Corinthians 5:1–4.

16. Ellul, *Prayer and Modern Man*, vii–viii.

17. Ibid.

18. Ibid., 174.

19. Jacques Ellul, *Politics of God and Politics of Man*, trans. by Geoffrey W. Bromiley (Grand Rapids: Eerdmans, 1972), 133–142 and 196–199 are good examples of Ellul's understanding of "efficacy" as I am describing it here.

20. Ibid., 174.

21. Ellul, *Judgment of Jonah*, 56–57.

22. Ibid.

23. Ellul, *Ethics of Freedom*, 127. The fuller discussion on prayer here on pages 124–128 of Ethics reflects what is referenced and commented on in this section.

24. Ellul, *Hope in Time of Abandonment*, 278.

25. Ellul, *Prayer and Modern Man*, vi.

26. Ibid., 174.

27. Ellul, *Ethics of Freedom*, 176–177; Ellul, *Living Faith*, 101–104 and 129–130; Ellul, *Money and Power*, 118; Ellul, *Judgment of Jonah*, 49–50; Ellul, *To Will and to Do*, 12 and 24–25; Ellul, *Prayer and Modern Man*, 124–125; Ellul, *Humiliation of the Word*, 95; Ellul, *Freedom, Love, and Power*, 78; Ellul, *Meaning of the City*, 15–16 and 48; Ellul, *Théologie et Technique*, 204; and Ellul, *Subversion of Christianity*, 24–26.

28. Ellul, *Living Faith*, 101–102. Ellul's emphasis.

29. Ibid.

30. See Ellul, *Hope in Time of Abandonment*, v–x for a description of this crisis and 89–97 for a discussion of "The silence of God."

31. Ellul, *Living Faith*, 101–104. Ellul, *Ethics of Freedom*, 176–177.

32. Ellul, *Living Faith*, 101–102. Ellul's emphasis and ellipses.

33. Ellul, *Prayer and Modern Man*, 174.

34. Romans 8:26–27. Ellul offers a reading of these verses in *Prayer and Modern Man*, 61–62: "Only when the Holy Spirit intercedes, and in a way which *cannot be*

expressed, that is, which transcends all verbalizing, all language, then is the prayer prayer, and it is a relationship with God. *Only* then 'he who searches the hearts of men knows' [. . .] 'what is the mind of the Spirit, because the Spirit intercedes . . . according to the will of God.' We are forced to the conclusion that prayer is a gift from God, and that its reality depends on him alone." The Psalms and Lamentations, of course, are also essential scriptural locations for the form and content of the people of God's prayer.

35. Ellul, *Meaning of the City*, 48.

36. Ellul, *To Will and to Do*, 19. Ellul writes this in reference to Solomon and Ecclesiastes and it is reaffirmed in James 1:5.

37. Ellul, *Hope in Time of Abandonment*, 201. Ellul, *Ethics of Freedom*, 154–155.

38. Ibid.

39. In Ellul, *Ethics of Freedom*, 52–59, Ellul reads the temptation to turn stones into bread as an analogy for our being tempted to rely on technology instead of the Word of God. In my dissertation, I show how this reading of the temptations of Jesus in the wilderness is inspired by Fyodor Dostoevsky's *Grand Inquisitor*, the story within the story *Brothers Karamazov*: Morelli, Michael Morelli, *Either Technological Pessimism or Proleptic Christian Resistance,* https://ethos.bl.uk/OrderDetails.do?uin=uk.bl.ethos.787402.

40. Matthew 4:4.

41. Ellul also writes about the church in the world in Ellul, *Ethics of Freedom*, 126–132. Here, Ellul engages with Karl Barth, CD I/2 and III/4, writing, "As a form of freedom the service of God, especially in the church and as distinct from service of our fellow-men, finds expression, as Barth tells us, in prayer, confession, the holy day, and reading scripture."

42. Ellul, *Theological Foundation of Law*, 128.

43. Ibid., 129. Ellul cites Ezekiel 33 in his reference to the church being a "watchman." Ellul cites Ezekiel 33 in his reference to the church being a "watchman." Watchman imagery is significant for Ellul, just as it is for Kierkegaard, who used the pseudonym *Vigilius Haufniensis*, which translates *Watchman of Copenhagen.*

44. Ibid., 130–131.

45. Ibid. Ellul cites Colossians 1:24 when he writes about the church being called to "[suffer] the same affiliations which Christ suffered for mankind."

46. Ibid., 133.

47. David Gill, *The Word of God in the Ethics of Jacques Ellul* (Metuchen: ATLA Monographs, 1984), 140–141 examines Ellul's understanding of the Holy Spirit and how this shapes his ethics. Gill correctly points out, "In a society ruled by technique, in politics as well as most other areas, the question of efficacy is extremely important. Ellul suggests that the effects of the Holy Spirit are of two types." First, "There are 'the mysterious spiritual effects which we cannot measure and which are found on earth and in heaven. These are totally [God's] (the birth of faith, the effectives of prayer) and will appear visibly at the end of the age.' " Second, the "qualification of Christian efficacy is, in fact, that it is always found in personal relationship with God in Jesus Christ" and "our responsibility is not for the total process but for our particular task." The textual references in these quotes are to Ellul's *Politics of*

God and Politics of Man, pages 135 and 138. These are helpful readings of Ellul's understanding of the Holy Spirit's presence and action in and through the church, so I hope to add to those readings here.

48. Ellul, *To Will and to Do*, 211.
49. Ibid., 213.
50. Ibid., 214.
51. Ibid., 220–221.
52. Ibid., 263.
53. Ibid.
54. Ibid., 265. See also Ellul, *Ethics of Freedom*, 270–274.
55. Ellul, *Ethics of Freedom*, 502–503.
56. Ellul, *Ethics of Freedom*, 503. "Cure/counseling" has been inserted in the paragraph to add depth to the translation. Whereas in this edition Bromiley uses "cure," "counseling" also could be used. Given Ellul's criticism of psychoanalysis and social psychology, "cure" should be read as "offering counsel," but in a biblical-theological sense (cff. Hebrews 4:12; Romans 15:14; and 2 Timothy 3:16–17) rather than a psychological/psychoanalytic sense. Similarly, on pages 253–254, Ellul writes about the "bad conscience of the pastor [spreading] to many members of the church" producing an "extraordinary desire to be exactly like other [people]" and to "want to be present everywhere, to have a part in all human activities, to be businessmen with businessmen, royalists with royalists, intellectuals with intellectuals, trade union men with trade union men. That is all. We no longer talk of conversion."
57. Ellul, *To Will and to Do*, 265.
58. Ephesians 4:11–16.
59. Ellul, *Ethics of Freedom*, 298.
60. Ibid., 502.
61. Ibid., 472.
62. Ibid.
63. Ellul, *To Will and to Do*, 265.
64. Ellul, *Ethics of Freedom*, 389 and 391. Ellul also develops an ethic of "non-power" in relation to technologies in Ellul, "Prolongements éthiques," *Théologie et Technique*, 307–352.
65. Ibid.
66. Jacques Ellul, *Les sources de l'éthique chrétienne: Le Vouloir et le Faire, parties IV et V* (*Sources of the Christian Ethic: To Will and to Do parts 4 and 5*), ed. Frédéric Rognon (Genève: Labor et Fides, 2018), 90. All capitalization of words in this passage are Ellul's. The passage is part of the chapter, "Une Éthique Commune," or "A Communal Ethic," which, along with the entirety of this text, could help compensate for Ellul's relative lack of a discussion of ecclesial community when compared to his focus on the individual in his theological and ethical writing.
67. Ellul, *Meaning of the City*, 176. Also, *False Presence of the Kingdom*, 21.
68. Ibid.
69. Ibid.

70. Jacques Ellul, *Apocalypse*, trans. George W. Schreiner (New York: Seabury Press, 1975). 222. Also, Ellul, *Ethics of Freedom*, 76–88 and Ellul, *What I Believe*, 188–210 and 214–233.

71. Ellul, *Meaning of the City*, 176. Also, *False Presence of the Kingdom*, 21. See also chapter three for brief discussion on opening and closing the dialectic of history in Ellul and Virilio's work.

72. Ellul, *Apocalypse*, 119. Ellul's emphasis.

73. Ellul, *Reason for Being*, 251.

74. Ellul, *Hope in Time of Abandonment,* 209.

75. Ellul, *Presence in the Modern World*, 59–66; Ellul, *To Will and to Do*, 25–29 and 218–222; and Ellul, *Hope in Time of Abandonment*, 258–282 are good examples.

76. Chris Huebner, "Patience, Witness, and the Scattered Body of Christ: Yoder and Virilio on Knowledge, Politics, and Speed," *A Precarious Peace* (Waterloo: Herald Press, 2006), 115–132 is an example. Huebner is an insightful reader of prominent themes in Virilio's work, and while he develops a nice comparative reading of Virilio and Yoder, he also unfairly argues, "Although Virilio is helpful in diagnosing the contemporary escalation of violence even in the absence of acts of war, he is noticeably less instructive in articulating the possible forms that resistance might take. When he is pushed on the question of what it would mean to resist the kind of totalizing violence he outlines, Virilio tends to fall back on typically banal liberal clichés that appeal to education and better understanding." As these sections will show, Virilio offers more resources than Huebner accounts for here. Similarly, Mark Lacy, *Security, Technology, and Global Politics: Thinking With Virilio* (London: Routledge, 2014), 149–150 is a good introduction to Virilio's work, and helpfully observes, "For the younger generation of radicals [Virilio] is too timid, offering revelation not revolution. But for Virilio what we need to hold on to is our capacity for negativity, our capacity to see what lies beneath the 'propaganda of progress.' " But, Lacy concludes, "The more profound hope in Virilio's world comes from his 'method,' his commitment to our capacity to keep asking questions about speed, control, power, and security in modernity." This is partially but not entirely correct. These sections will show how there is more hope to be found in Virilio's writing, beyond just a questioning of speed, control, power, and security.

77. Virilio and Lotringer, *Crepuscular Dawn*, 147–148 and Virilio and Lotringer, *The Accident of Art*, 88, which seems to feature content taken directly from *Crepuscular Dawn*. See also, Virilio, *Grey Ecology*, 52–53 and Virilio and Armitage, *Virilio Live*, 20–21. Ellul often says the same. See, for example, Ellul, *Humiliation of the Word*, 268: to people asking for strictly positivistic solutions, "I could answer in learned fashion with the positivistic dialectic of negativism: in the final analysis, only the No produces the change and advance." As such, "Whoever accuses my analyses and research of negativism and considers iconoclasm and criticism of structuralist ideologies or of romanticizing insanity to be a purely pessimistic orientation proves just one thing: that he loves his chains."

78. Virilio, *Lost Dimension*, 109–110. Virilio's emphasis.

79. Ibid., 38–39.

80. Virilio and Armitage, *Virilio Live*, 20–21.

81. Virilio, *War and Cinema*, 58–59 and 81–82; *Lost Dimension*, 127; *Negative Horizon*, 182; *Ground Zero*, 40; and *University of Disaster*, 59. The quote is from Dietrich Bonhoeffer, *Discipleship*, trans. Barbara Green and Reinhard Krauss (Minneapolis: Fortress Press, 2003), 94–95.

82. Bonhoeffer, *Discipleship*, 93–94. Bonhoeffer's emphasis.

83. Ibid., 95.

84. Virilio, *City of Panic*, 56. Author's emphases. As mentioned earlier, note here the similarities between Virilio's descriptions of "the synchronization of individual emotions" and Ellul's descriptions of the "democratization of evil." For other examples of Virilio's negatively rendered Christological imagery, see Virilio, *Aesthetics of Disappearance*, 38; Virilio, *The Vision Machine*, 16; Virilio, *Politics of the Very Worst*, 41; Virilio, *Ground Zero*, 10–12; Virilio, *The Original Accident*, 45 which is prefaced by Luke 23:34; Virilio, *Grey Ecology*, 76; Virilio, *The Great Accelerator*, 87.

85. For exactly these reasons, chapters like Virilio, "Kreigstrasse," *City of Panic*, 47–61; Virilio, "L'État suicidaire" and "L'évangile nucléaire," *L'insécurité du territoire*, 25–52 and 113–141; and "IV," *University of Disaster*, 95–117 come alive, theologically, ethically, and aesthetically when read with traditional two cities doctrine in mind.

86. Virilio, *Crepuscular Dawn*, 135–136. Virilio's emphasis.

87. Virilio, *War and Cinema*, image 42, in the chapter "Traveling Shot."

88. Virilio, *University of Disaster*, 118–199. See also Virilio, *Ground Zero*, 19–20 and Virilio, *Administration of* Fear, 81–82 for Virilio's account of the Trinity Site and Trinity Test. For other examples of Trinitarian imagery in Virilio's writing, similar to what is referenced here, see Virilio, *Desert Screen*, 4–5; Virilio, *City of Panic*, 124; Virilio and Lotringer, *Pure War*, 195–196; and Virilio and Armitage, *Virilio Live*, 86.

89. Virilio and Brausch, *A Winter's Journey*, 102. For other places where Virilio offers a positive rendering of his understanding of Jesus Christ, see Virilio and Lotringer, *Crepuscular Dawn*, 62 and Virilio and Lotringer, *The Accident of Art*, 101.

90. Bonhoeffer, *Discipleship*, 93–94. Bonhoeffer's emphasis.

91. Virilio, *Negative Horizon*, 27. See *Aesthetics of Disappearance*, 52 for another example in which Virilio describes the mystical visions of Saint Bernadette Soubirous and calls her a worthy contemporary of the German idealists. Or Virilio, *City of Panic*, 129–132 in which Virilio cites "The Canticle of All Creatures" by Saint Francis of Assisi, quotes Saint Hildegard von Bingen—"Man is the enclosure of God's marvels"—and works with theological imagery in relation to the desert. As quoted earlier, see Virilio, *University of Disaster*, 134–135 for his reference to the Psalmist proclaiming that God's truth will spring up from the ground.

92. Ellul, *Hope in Time of Abandonment*, 89–97 and 3–9.

93. Virilio, *L'insécurité du territoire*, 167. For glimpses of Virilio's understanding of eschatology see Virilio, *Aesthetics of Disappearance*, 18 and 47–48. Virilio, *University of* Disaster, 15. Virilio, *The Art of the Motor*, 68. Virilio and Lotringer, *Crepuscular Dawn*, 158–159. Virilio, *Grey Ecology*, 43.

94. Virilio, *L'insécurité du territoire*, 167–168.

95. Virilio, *University of Disaster*, 28.

96. Virilio, *The Great Accelerator*, 89. Virilio's emphasis.

97. Virilio, *The Vision Machine*, 17. See also Virilio, *Landscape of Events*, 66 for similar, though less explicit, themes and imagery.

98. Bonhoeffer, *Ethics*, 399–400. Bonhoeffer's emphasis.

99. Bonhoeffer, *Creation and Fall*, 112–113. Virilio, *Ground Zero*, 9–10 references *Creation and Fall* in Virilio's discussion of "the **sicut deus** of the Luciferian bargain" as it relates to theology, ethics, and aesthetics. Virilio's emphasis. See also Virilio, *Lost Dimension*, 109–110.

100. Virilio and Brausch, *A Winter's Journey*, 102. For other places where Virilio offers a positive rendering of his understanding of Jesus Christ, see Virilio and Lotringer, *Crepuscular Dawn*, 62 and Virilio and Lotringer, *The Accident of Art*, 101.

101. Virilio, *The Accident of Art*, 35. Virilio's emphasis. While the text says Virilio references Psalm 18 it actually seems he means Psalm 19. The reasons for this possible discrepancy are unclear, but suffice it to say, Psalm 19 better captures what Virilio is speaking about here. Also, the reference to André Malraux is to Malraux's book *Voices of Silence: Man and His Art*.

102. Virilio, *Lost Dimension*, 115–116. The motifs of creation, time, rhythm, night and day, rest, and Sabbath frequently appear in Virilio's writing. See, for example, The *Vision Machine*, 26; Virilio and Lotringer, *Pure War*, 48 and 140; Virilio, *University of Disaster*, 124; Virilio and Richard, 44–45; Virilio, *The Great Accelerator*, 31–32.

103. Ellul also writes about the importance of Sabbath for the church. See Ellul, *Ethics of Freedom*, 128–130 and Ellul, *Théologie et Technique*, 206–215. In the latter case Ellul writes at length about how Sabbath places a limit on, and resists, the technological system.

104. Virilio, *Bunker Archaeology*, 14. Virilio's elipsis.

105. Ibid., 46.

106. Virilio and Lotringer, *Crepuscular Dawn*, 24.

107. Ibid., 24.

108. Ibid., 28.

109. Matthew 28 and John 30.

110. Virilio and Armitage, *Virilio Live*, 174–175. In the mid-nineteenth century, at fourteen years old, Saint Bernadette received multiple visions of the Virgin Mary in Lourdes, France. In one of her visions, she was asked to build a chapel at a nearby cave-grotto, which at the time, was a repository for garbage and a common meeting point for prostitutes and thieves. Now the cave-grotto is a holy site where miraculous healings and visions have been reported. Virilio talks or writes about this in such places as Virilio and Lotringer, *Crepuscular Dawn*, 26–28 and Virilio, *Aesthetics of Disappearance*, 52–53.

111. Virilio, *L'insécurité du territoire*, 136. Virilio writes run-on sentences in the French edition, and while it would be tempting to break the sentences up after translating them to make reading this paragraph easier, leaving Virilio's writing as it is captures the forceful style of his prose. The essay is called "L'évangile nucléaire," or, "The Nuclear Gospel," 113–141. It is important because it offers Virilio's

most detailed account of a theology of church and state. Specifically, he reframes Augustine's city of man/City of God theology, by bringing it into the socio-historical context of the Cold War nuclear arms race. This cannot be discussed at length here, but it adds theological nuance to Virilio's descriptions of church/state relations, particularly as they relate to the Eucharist and the Crucifixion, in this passage. As mentioned earlier, see Virilio, "Kreigstrasse," *City of Panic*, 47–61; Virilio, "L'État suicidaire" and "L'évangile nucléaire," *L'insécurité du territoire*, 25–52 and 113–141; and "IV," *University of Disaster*, 95–117 for other Virilian renderings of traditional two cities doctrine.

112. This is most obvious when Virilio references Nicholas of Cusa. For example, Virilio, "The Third Interval: A Critical Transition," *Rethinking Technologies*, 9. This appears as a chapter in Virilio, *Open Sky*, 9–21. Virilio does not state directly that he affirms a theology of transubstantiation, but given he is Catholic, and in light of the passages cited and themes explored in this section, it is likely he does. Whether it is a theology of transubstantiation, real presence, or symbolism, the Eucharist, at most, actually manifests the unification of accident and substance, or at least, symbolizes the unification of accident and substance. In any case, it is an essential sacramental basis for the church's and believer's sensitization to the relationships between the accidents and substances of creation.

113. Frances Stonor Saunders, *The Cultural Cold War*, 5. Authors emphasis.

114. Ibid.

115. Ellul, *Propaganda*, 111. Ellul's emphasis.

116. Ellul, *New Demons*, 113.

117. Ephesians 4:14.

118. Sharon Weinberger, *The Imagineers of War* (New York: Alfred A. Knopf, 2017), 255–256.

119. Virilio, *The Information Bomb*, 109–110.

120. On this subject, Florian Rötzer, *Conversations with French Philosophers*, trans. Gary E. Aylesworth (New Jersey: Humanities Press, 1995) helpfully explains, through introductory essays and interviews with thinkers like Baudrillard, Derrida, Lyotard, Serres, as well as Virilio, why these prominent twentieth-century scholars write in a way that seems so irrational and unreasonable to those unfamiliar with their methodologies and styles. Virilio's interview is on 97–104, in which Virilio opens the interview saying, "I claim for myself this status as an outsider and also this ghetto, because France is still the paradise of cartesianism. That goes for morals and also for relations between human beings. Despite the events of [1968] and despite a certain change of consciousness, the atmosphere is still dominated by rationalism and even uncleansed by a rationalism of the worst sort." It offers a good explanation for why thinkers like Virilio and Ellul write in the way they do.

121. Of course, it has always been thus. However, after Francis Fukuyama, the narrative of the end of history has prevailed for a generation, and events as disparate as Brexit; President Donald Trump's ascendancy to and fall from power, and the subsequent storming of the Capitol Building in Washington; Occupy Wall Street; the Yellow Jacket protests in France; the Charlottesville revolts in Virginia; the killings of George Floyd and Breonna Taylor and the protests that followed; and, a proverbial

host of other events taking place on the world stage show how the sociopolitical, and with it the moral, center is no longer holding. One of Ellul's first articles, "Fascism, Son of Liberalism," *Political Illusion and Reality*, eds. David Gill and David Lovekin, trans. Jacob Rollison (Eugene: Wipf and Stock, 2018), 3–45 comes to mind here.

122. Matthew 16:26 and Mark 8:36. See Ellul, *Reason for Being*, 90–91 and Virilio, *Open Sky*, 25 and 64 for places in which both authors deploy these verses.

123. Matthew 16:24–25 and Mark 8:34–35.

124. Note, I am not altogether opposed to cyborgs. In fact, I have benefited from technologies, which, from a certain perspective, could lead people to designate me a cyborg. I wear glasses, for instance, and I regularly use medical supplies, which, based on my understanding of cyborgism, could make me a cyborg. Here, though, I am trying to emphasize and maintain the power of Paul's image of the blood, bone, and flesh Body of Christ (i.e., the Incarnated body of Christ and the incarnate church) to resist an image of the church, which affixes manipulative and harmful, but unnecessary, technologies to this actual body.

125. 1 Corinthians 1:18.

126. 1 Corinthians 6:19–20.

127. Matthew 25:31–46 for fuller context.

128. Matthew 18:2–10; Mark 10:13–16; Luke 9:40–44; and John 5:5–15.

129. Of course, not only in discussions of power. There must be inclusion in *all* discussions.

Bibliography

Alonsi, Adoni. "An Unreasonable Thinker: How Ellul Engages Cybercultural Criticism." In *Jacques Ellul and the Technological Society in the 21st Century: Philosophy of Engineering and Technology* 13 (2013), edited by Helena M. Jerónimo, José Luís Garcia and Carl Mitcham. New York: Springer, 2013, 115–128.

Alperovitz, Gar and Martin J. Sherwin. "Op-Ed: U.S. Leaders Knew We Didn'T Have to Drop Atomic Bombs on Japan to Win the War. We Did It Anyway." *Los Angeles Times*, August 5, 2020. https://www.latimes.com/opinion/story/2020-08-05/hiroshima-anniversary-japan-atomic-bombs.

Arnould, Jacques. *L'Apocalypse du vide*. Paris: Edilivre, 2019.

Augustine. *City of God Against the Pagans*. Translated by Philip Levine. Cambridge: Harvard University Press, 1966.

Barth, Karl. "The Problem of Ethics Today." In *The Word of God and The Word of Man*, translated by Douglas Horton. London: Hodder and Stoughton, 1928.

Beggin, Riley. "Trump Threatens New Attacks on Iran, as Iranian Leaders Compare Him to ISIS." *Vox*, January 5, 2020. https://www.vox.com/world/2020/1/5/21050347/trump-new-attacks-iran-soleimani-52-military-cultural .

Benjamin, Walter. *Illuminations*. Edited by Hannah Arendt. Translated by Harcourt Brace Jovanovich. New York: First Mariner Books, 2019.

Bonhoeffer, Dietrich. *Creation and Fall*. Edited by John W. De Grucy. Translated by Douglas Stephen Bax. Minneapolis: Fortress Press, 1997.

Bonhoeffer, Dietrich. *Discipleship*. Translated by Barbara Green and Reinhard Krauss. Minneapolis: Fortress Press, 2003.

Bonhoeffer, Dietrich. *Ethics*. Edited by Clifford J. Green. Translated by Reinhard Krauss, Charles C. West and Douglas W. Stott. Minneapolis: Fortress Press, 2009.

Brock, Brian. "Culture as Flight From God." In *Fall Narratives*, edited by Zohar Hadromi-Allouce and Áine Larkin. London: Routledge, 2017.

Burdett, Michael. *Eschatology and the Technological Future*. New York: Routledge, 2015.

Chesnaux, Jean. *De la modernité*. Paris: Maspero, 1983.

Clendenin, Daniel. *Theological Method in Jacques Ellul.* New York: University Press of America, 1987.

Cole, Matthew and Sheelah McNeill. "American Sniper" Chris Kyle Distorted His Military Record, Documents Show." *The Intercept*, May 25, 2016.https://theintercept.com/2016/05/25/american-sniper-chris-kyle-distorted-his-military-record-documents-show/.

Dawn, Marva. *The Concept of "The Principalities and Powers" in Jacques Ellul.* Notre Dame: PhD Dissertation, 1992.

de Grazia, Victoria. *Irresistible Empire: America's Advance through 20th-Century Europe.* London: Belknap Press, 2005.

Descartes, René. *Meditations.* Translated by Elizabeth S. Haldane and G. Ross. Chicago: Encyclopaedia Britannica, 1952.

Duriez, Pierre. *Correspondence with Jacques Ellul.* Manuscrits des universités de Bordeaux. Manuscrits de Sciences Po Bordeaux. Carton 1/117. December 1, 1992.

Ellul, Jacques. "La Bible et la ville." *Foi et Vie* 48, no. 1 (janvier-février 1950): 4–19.

Ellul, Jacques. "Mythes Modernes." *Diogène* 6, no. 23 (September 1958): 23–40.

Ellul, Jacques. *Theological Foundation of Law.* Translated by Marguerite Wiser. New York: Seabury Press, 1960.

Ellul, Jacques. *False Presence of The Kingdom.* Translated by C. Edward Hopkin. New York: Seabury Press, 1963.

Ellul, Jacques. *The Technological Society.* Translated by John Wilkinson. New York: Vintage Books, 1964.

Ellul, Jacques. *Political Illusion.* Translated by Konrad Kellen. New York: Vintage Books, 1967.

Ellul, Jacques. *Histoire de la Propagande.* Paris: Presses Universitaires de France, 1967.

Ellul, Jacques. *A Critique of the New Commonplaces.* Translated by Helen Weaver. New York: Alfred A. Knopf, 1968.

Ellul, Jacques. *To Will and To Do.* Translated by C. Edward Hopkin. Philadelphia: Pilgrim Press, 1969.

Ellul, Jacques. *Violence: Reflections From a Christian Perspective.* Translated by Cecelia Gaul Kings. New York: Seabury Press, 1969.

Ellul, Jacques. "Cain, the Theologian of 1969." *Katallagete: Be Reconciled* 2, Winter (1968–1969): 4–7.

Ellul, Jacques. *Autopsy of Revolution.* Translated by Patricia Wolf. New York: Alfred A. Knopf 1971.

Ellul, Jacques. *Politics of God and Politics of Man.* Translated by Geoffrey W. Bromiley. Grand Rapids: Eerdmans, 1972.

Ellul, Jacques. *Hope in Time of Abandonment.* Translated by C. Edward Hopkin. New York: Seabury Press, 1973.

Ellul, Jacques. *Prayer and Modern Man.* Translated by C. Edward Hopkin. New York: Seabury Press, 1973.

Ellul, Jacques. *Propaganda.* Translated by Konrad Kellen and Jean Lerner. New York: Vintage Books, 1975.

Ellul, Jacques. *The New Demons.* Translated by C. Edward Hopkin. New York: Seabury Press, 1975.

Ellul, Jacques. *Apocalypse*. Translated by George W. Schreiner. New York: Seabury Press, 1975.

Ellul, Jacques. *The Ethics of Freedom*. Translated by Geoffrey W. Bromiley. London: Mowbrays, 1976.

Ellul, Jacques. *The Technological System*. Translated by Joachim Neugroschel. New York: Continuum, 1977.

Ellul, Jacques. *The Betrayal of the West*. Translated by Matthew J. O'Connell. New York: Seabury Press, 1978.

Ellul, Jacques. *Living Faith*. Translated by Peter Heinegg. San Francisco: Harper and Row Publishers, 1980.

Ellul, Jacques. "On Dialectic." In *Jacques Ellul: Interpretative Essays*, edited by Clifford G. Christians and Jay M. Van Hook, translated by Geoffrey W. Bromiley. Chicago: University of Illinois Press, 1981.

Ellul, Jacques. "La democratisation du mal." *Sud-Ouest*, September 6, 1981.

Ellul, Jacques and Madeleine Garrigou-Lagrange. *In Season out of Season*. Translated by Lani K. Niles. Cambridge: Harper and Row, 1982.

Ellul, Jacques. *Humiliation of the Word*. Translated by Joyce Main Hanks. Grand Rapids: Eerdmans, 1985.

Ellul, Jacques. *La Genèse aujourd'hui*. Paris: Editions de L'AREFPPI, 1987.

Ellul, Jacques. *What I Believe*. Translated by Geoffrey W. Bromiley. Grand Rapids: Eerdmans, 1989.

Ellul, Jacques. *Reason for Being*. Translated by Joyce Main Hanks. Grand Rapids: Eerdmans, 1990.

Ellul, Jacques. *The Technological Bluff*. Translated by Geoffrey W. Bromiley. Grand Rapids: Eerdmans, 1990.

Ellul, Jacques. "The Contemporaneity of the Reformation." *Sources and Trajectories*. Translated by Marva Dawn. Grand Rapids: Eerdmans, 1997.

Ellul, Jacques. *Subversion of Christianity*. Translated by Geoffrey W. Bromiley. Eugene: Wipf and Stock, 2001.

Ellul, Jacques. *Perspectives on our Age*. Edited by Willem H. Vanderburg. Translated by Canadian Broadcasting Corporation. Toronto: House of Anansi Press, 2004.

Ellul, Jacques and Patrick Troude-Chastenet. *Jacques Ellul on Politics, Technology, and Christianity*. Translated by Joan Mendès France. Eugene: Wipf and Stock 2005.

Ellul, Jacques. *Freedom, Love, and Power*. Translated by Willem Vanderburg. Toronto: University of Toronto Press, 2010.

Ellul, Jacques. *Judgment of Jonah*. Translated by Geoffrey W. Bromiley, Eugene: Wipf and Stock, 2011.

Ellul, Jacques. *Meaning of The City*. Translated by Denis Pardee. Eugene: Wipf and Stock, 2011.

Ellul, Jacques. *Le Bluff Technologique*. Paris: Pluriel, 2012.

Ellul, Jacques. *Empire of Non-Sense*. Translated by Michael Johnson and David Lovekin. Berkshire: Papadakis, 2014.

Ellul, Jacques. *Théologie et Technique: Pour une éthique de la non-puissance*. Edited by Yves Ellul and Frédéric Rognon. Genève: Labor et Fides, 2014.

Ellul, Jacques. *Presence in The Modern World.* Translated by Lisa Richmond. Eugene: Cascade Books, 2016.

Ellul, Jacques. *Histoire des institutions: L'Antiquité.* Paris: PUF, 2016.

Ellul, Jacques. "Current Signification of the Reformation." *European Journal of Theology* xxvi, no.2 (2017). Translated by Jacob Rollison.

Ellul, Jacques. *La Pensée Marxiste.* Edited by Michel Hourcade, Jean-Pierre Jezequel and Gérard Paul. Paris: La Table Ronde, 2018.

Ellul, Jacques. *Les sources de l'éthique chrétienne: Le Vouloir et le Faire, parties IV et V.* Edited by Frédéric Rognon. Genève: Labor et Fides, 2018.

Ellul, Jacques. "Fascism, Son of Liberalism." In *Political Illusion and Reality*, edited by David Gill and David Lovekin. Tanslated by Jacob Marques-Rollison. Eugene: Wipf and Stock, 2018.

Ellul, Jérôme. Interview by Michael Morelli. Conversation. Vancouver, June 28, 2018.

Erickson, Amy. *Ephraim Radner, Hosean Wilderness, and the Church in the Post-Christendom West: A Dialogue on the Shape of Waiting.* Leiden: Brill, 2020.

Fasching, Darrell. *The Thought of Jacques Ellul.* Toronto: Edwin Mellen Press, 1981.

Fasching, Darell. "The Sacred, the Secular, and the Holy." *Ellul Forum* no. 43 (April 2014), https://journals.wheaton.edu/index.php/ellul/article/download/414/16/.

Foucault, Michel. *The Archaeology of Knowledge and Discourse on Language.* Translated by A.M. Sheridan Smithi. New York: Routledge, 2002.

Gill, David. *The Word of God in the Ethics of Jacques Ellul.* Metuchen: ATLA Monographs, 1984.

Goddard, Andrew. *Living the Word, Resisting the World: The Life and Thought of Jacques Ellul.* Carlisle: Paternoster Press, 2002.

Greeman, Jeffrey, Read Schuchardt and Noah Toly. *Understanding Jacques Ellul.* Eugene: Cascade Books, 2012.

Halmi, Nicholas. *The Genealogy of the Romantic Symbol.* Oxford: Oxford University Press, 2007.

Hawkins, Derk. "Brian Williams is 'Guided by the Beauty of Our Weapons' in Syria Strikes." *Washington Post*, April 7, 2017. https://www.washingtonpost.com/news/morning-mix/wp/2017/04/07/beautiful-brian-williams-says-of-syria-missile-strike-proceeds-to-quote-leonard-cohen/.

Heidegger, Martin. *The Question Concerning Technology.* Translated by William Lovitt. New York: Harper Perennial, 2013.

Heubner, Chris. *A Precarious Peace.* Waterloo: Herald Press, 2006.

Hijaya, James A. "The *Gita* of Robert Oppenheimer." *Proceedings of the American Philosophical Society* 144, no. 2 (June 2000): 123–167.

Hirsch, Michael. "Complexity in the Age of Simplicity." In *Grey Ecology*, translated by Drew Burk. New York: Atropos Press, 2009.

"Homo faber, n." OED Online. March 2019. *Oxford University Press.* http://www.oed.com/view/Entry/272606?redirectedFrom=homo+faber&.

James, Ian. *Paul Virilio.* New York: Routledge, 2007.

Janicaud, Dominique. *Powers of the Rational.* Translated by Peg and Elizabeth Birmingham. Bloomington: Indiana University Press, 1994.

Judt, Tony. *Postwar: A History of Europe Since 1945*. London: Vintage Books, 2010.

Kellner, Douglas. "Virilio, War, and Technology." In *From Modernism to Hypermodernism and Beyond*, edited by John Armitage. London: Sage, 2000.

Lacy, Mark. *Security, Technology, and Global Politics: Thinking With Virilio.* London: Routledge, 2014.

Lacy, Mark. "Paul Virilio (1932–2018)." *Verso*, September 20, 2018. https://www.versobooks.com/blogs/4042-paul-virilio-1932-2018.

Leroi-Gourhan, André. *Evolution et techniques, vol. 2: Milieu et techniques.* Paris: Albin Michel, 1945.

Lovekin, David. *Technique, Discourse, and Consciousness.* Bethlehem: Leigh University Press, 1991.

Lovekin, David. "Art and Technology: All or Nothing." In *Empire of Non-Sense.* Edited and translated by David Lovekin and Michael Johnson. Berkshire: Papadakis, 2014.

Luther, Martin. *Luther's Works Volume 1: Lectures on Genesis Chapters 1–5.* Edited and translated by Jaroslav Pelika. Saint Louis: Concordia Publishing House, 1968.

Lyle, Emily B. "Dumézil's Three Functions and Indo-European Cosmic Structure." *History of Religions* 22, no. 1 (1982): 25–44.

Marx, Karl and Frederick Engels. *Marx/Engels Selected Works*, volume 1. Translated by Samuel Moore and Frederick Engels. Moscow: Progress Publishers, 1969.

Marx, Karl and Fredrick Engels. *Marx and Engels Collected Works Volume 35: Capital Vol. I.* London: Lawrence and Wishart, 1996.

Mellville, Herman. "A Canticle." In *Battle-Pieces and Aspects of the War.* New York: Thomas Yoseloff, 1964.

Morelli, Michael. *Either Technological Pessimism or Proleptic Christian Resistance: A Constructive Reading of Jacques Ellul's Dialectical Writings on Technology with Paul Virilio as an Evocative Interlocutor.* University of Aberdeen, Aberdeen, PhD Dissertation, 2019. https://ethos.bl.uk/OrderDetails.do?uin=uk.bl.ethos.787402.

Morelli, Michael. "What's in a Name? Jacques Ellul's Reading of Naming in Genesis 1–3." In *Jacques Ellul and the The Bible: Towards a Hermeneutic of Freedom.* Eugene: Pickwick Publications, 2020.

Reinert, Hugo and Erik Reinert. "Creative Destruction in Economics: Nietzsche, Sombart, and Schumpeter." *The European Heritage in Economics and the Social Sciences* 3 (2006): 55–85.

Rieger, Bernhard. *The People's Car: A Global History of the Volkswagen Beetle.* Cambridge: Harvard University Press, 2013.

Rose, Julie. "Translators Preface." In *Art and Fear*, translated by Julie Rose. London: Continuum, 2000.

Rötzer, Florian. *Conversations with French Philosophers.* Translated by Gary E. Aylesworth. New Jersey: Humanities Press, 1995.

Seyhan, Azade. "What is Romanticism, and Where Did It Come From?" In *The Cambridge Companion to German Romanticism*, edited by Nicholas Saul. Cambridge: Cambridge University Press, 2009.

Sokal, Alan and Jean Bricmont. "Paul Virilio." In *Fashionable Nonsense: Postmodern Intellectuals' Abuse of Science*. New York: Picador, 1998.

Stearns, Peter N. "Pinpointing the Industrial Revolution: Features, Times, and Place." In *Debating the Industrial Revolution*. New York: Bloomsbury Publishing, 2015.

Stoke, Bruce. "70 Years after Hiroshima, Opinions have Shifted on Use of Atomic Bomb." *Pew Research Centre*, August 4, 2015. https://www.pewresearch.org/fact-tank/2015/08/04/70-years-after-hiroshima-opinions-have-shifted-on-use-of-atomic-bomb/.

Stonor Saunders, Frances. *The Cultural Cold War: The CIA and The World of Arts and Letters*. New York: The New Press, 1999.

Temple, Katherine. *The Task of Jacques Ellul*. Toronto: McMaster University PhD Dissertation, 1976.

Thielman, Sam. "How Do You Stop the Far-Right Using the Punisher Skull?" *The Guardian*, June 11, 2020. https://www.theguardian.com/books/2020/jun/11/how-do-you-stop-the-far-right-using-the-punisher-skull-make-it-a-black-lives-matter-symbol.

Turnbull, Neil. "Light and Illumination: Paul Virilio's Neoplatonism and Theological Critique of Modernity." *Cultural Politics* 11, no. 2 (2015): 260–274.

Van Vleet, Jacob. *Dialectical Theology and Jacques Ellul*. Minneapolis: Fortress Press, 2014.

Violeau, Jean-Louis. "Introduction: Critical Space Under Nuclear Winter." In *Lost Dimension*, translated by Daniel Moshenberg. Los Angeles: Semiotexte 1984.

Virilio, Paul. *Desert Screen*. Translated by Michael Degener. London: Continuum, 1981.

Virilio, Paul. *War and Cinema*. Translated by Patrick Camiller. London: Verso, 1984.

Virilio, Paul. *Lost Dimension*. Translated by Daniel Moshenberg. Los Angeles: Semiotexte, 1984.

Virilio, Paul. *Negative Horizon*. Translated by Michael Degener. New York: Continuum, 1984.

Virilio, Paul and Frédéric Gaussen. "Violence, Ralentir." *Le Monde*, June, 1986.

Virilio, Paul. *The Vision Machine*. Translated by Julie Rose. Bloomington: BFI Publishing, 1988.

Virilio, Paul. *Polar Inertia*. Translated by Patrick Camiller. London: SAGE Publications, 1990.

Virilio, Paul. *Popular Defense and Ecological Struggles*. Translated by Mark Polizzotti. New York: Semiotexte, 1990.

Virilio, Paul and Mark Hunyadi. "Vitesse, pouvoir absolu."*Journal de Genève et Gazette de Lausanne* 268 (November 14, 1992): 22.

Virilio, Paul. *L'insécurité du territoire*. Paris: Éditions Galilée, 1993.

Virilio, Paul. *The Art of The Motor*. Translated by Julie Rose. Minneapolis: University of Minnesota Press, 1993.

Virilio, Paul. "The Third Interval: A Critical Transition." In *Rethinking Technologies*, edited by Verna Andermatt Conley, translated by Tom Conley. Minneapolis: University of Minnesota Press, 1993.

Virilio, Paul. *Bunker Archaeology*. Translated by George Collins. New York: Princeton Architectural Press, 1994.

Virilio, Paul. *Open Sky*. Translated by Julie Rose. London: Verso Books, 1995.

Virilio, Paul. *Landscape of Events*. Translated by Julie Rose. Cambridge: MIT Press, 1996.
Virilio, Virilio and Philippe Petit. *Politics of the Very Worst*. Translated by Michael Cavaliere. New York: Semiotexte, 1996.
Virilio, Paul and Marianne Brausch. *A Winter's Journey*. Translated by Chris Turner. London: Seagull Books, 1997.
Virilio, Paul. *The Information Bomb*. Translated by Chris Turner. London: Verso, 1998.
Virilio, Paul. *Strategy of Deception*. Translated by Chris Turner. London: Verso, 1999.
Virilio, Paul. *Art and Fear*. Translated by Julie Rose. London: Continuum, 2000.
Virilio, Paul and John Armitage. *Virilio Live: Selected Interviews*. London: SAGE Publications 2001.
Virilio, Paul. *Ground Zero*. Translated by Christ Turner. London: Verso, 2002.
Virilio, Paul. *City of Panic*. Translated by Julie Rose. Oxford: BERG, 2004.
Virilio, Paul. *The Original Accident*. Translated by Julie Rose. Cambridge: Polity Press, 2005.
Virilio, Paul and Sylvère Lotringer. *The Accident of Art*. Translated by Michael Taormina. New York: Semiotexte, 2005.
Virilio, Paul. *Art As Far As The Eye Can See*. Translated by Julie Rose. London: Bloomsbury, 2005.
Virilio, Paul. *Speed and Politics: An Essay on Dromology*. Translated by Mark Polizzotti. Los Angeles: Semiotexte, 2006.
Virilio, Paul. *The University of Disaster*. Translated by Julie Rose. Cambridge: Polity Press, 2007.
Virilio, Paul. *Grey Ecology*. Translated by Drew Burk. New York: Atropos Press, 2009.
Virilio, Paul. *The Futurism of the Instant*. Translated by Julie Rose. Cambridge: Polity Press, 2009.
Virilio, Paul. *The Great Accelerator*. Cambridge: Polity Press, 2010.
Virilio, Paul and Betrand Richard. *The Administration of Fear*. Translated by Amos Hodges. Los Angeles: Semiotexte, 2007.
Virilio, Paul and Sylvère Lotringer. *Pure War*. Translated by Mark Polizzoti. Los Angeles: Semiotexte, 2008.
Wagenfuhr, Gregory. *Plundering Egypt*. Eugene: Cascade Books, 2016.
Weinberger, Sharon. *The Imagineers of War*. New York: Alfred A. Knopf, 2017.

Index

About the Author

Michael Morelli (PhD, University of Aberdeen) is assistant dean and assistant professor of Undergraduate Studies and assistant professor of Theology, Culture and Ethics at Northwest College and Seminary, which is, a member of Associated Canadian Theological Schools and an affiliate of Trinity Western University.

www.ingramcontent.com/pod-product-compliance
Ingram Content Group UK Ltd.
Pitfield, Milton Keynes, MK11 3LW, UK
UKHW012119150926
475560UK00002B/178

9 781793 625434